T0137645

THERMAL AND POWER MANAGEMENT OF INTEGRATED CIRCUITS

SERIES ON INTEGRATED CIRCUITS AND SYSTEMS

Anantha Chandrakasan, Editor
Massachusetts Institute of Technology
Cambridge, Massachusetts, USA

Published books in the series:

A Practical Guide for SystemVerilog Assertions
Srikanth Vijayaraghavan and Meyyappan Ramanathan
2005, ISBN 0-387-26049-8

Statistical Analysis and Optimization for VLSI: Timing and Power
Ashish Srivastava, Dennis Sylvester and David Blaauw
2005, ISBN 0-387-25738-1

Leakage in Nanometer CMOS Technologies
Siva G. Narendra and Anantha Chandrakasan
2005, ISBN 0-387-25737-3

Thermal and Power Management of Integrated Circuits
Arman Vassighi and Manjo Sachdev
2005, ISBN 0-398-25762-4

THERMAL AND POWER MANAGEMENT OF INTEGRATED CIRCUITS

Arman Vassighi and Manoj Sachdev
Electrical and Computer Engineering, University of Waterloo, Waterloo, Canada

 Springer

Arman Vassighi
Department of Electrical and
  Computer Engineering
University of Waterloo
200 University Avenue, West
Waterloo N2L 3G1
Canada.

Manoj Sachdev
Department of Electrical and
  Computer Engineering
University of Waterloo
200 University Avenue, West
Waterloo N2L 3G1
Canada

Thermal and Power Management of Integrated Circuits

ISBN-10: 0-387-29749-9 (e-book)
ISBN-13: 9780387297491 (e-book)

ISBN: 978-1-4419-3832-9

Printed on acid-free paper.

9 8 7 6 5 4 3 2 1

springer.com

Contents

Preface

Our capability to integrate ever larger number of transistors is surprising even to the most ardent believers of scaling. At 2005 International Solid State Conference, Intel announced a processor with 1.7 billion transistors. This trend is likely to continue at least for a decade. However, there are a number of issues that must be dealt with if this integration trend is to continue. In this context, thermal and power management of ultra large scale integration (ULSI) is one of the major concerns.

Thermal issues are a bi-product of scaling and quest for speed. This issue came to the forefront as we scale the technology in nano-metric regime. However, it is not limited to high speed circuits alone. Today, even moderate speed ULSI must worry about containing the junction temperature under limits. The junction temperature affects large number of important device parameters, and an unabated increase in the junction temperature may have disastrous implications on performance, and long term reliability of integrated circuits. Moreover, manufacturing and operational costs such as expensive cooling solutions may increase significantly, if rising junction temperatures are not contained.

Recently, a great deal of attention has been paid to thermal issues and there are international conferences dedicated to thermal issues in electronics. However, lack of books in this area is impeding the awareness of the subject to engineers and managers who are suddenly confronted with these issues. This book makes an attempt to provide a comprehensive overview of the power and thermal management of integrated circuits.

Acknowledgement

We would like to express our gratitude to researchers at circuit research lab at Intel, Hillsboro for suggesting thermal issues as a research topic. In particular, we are indebted to Shekhar Borkar, Vivek De, and Ali Keshavarzi for their feedback and support for the research. We are also grateful to Oleg Semenov (University of Waterloo) and Chuck Hawkins (University of New Mexico) for wide ranging discussions on the topic.

We are thankful to Carl Harris of Springer for facilitating the publishing of this book. Finally, authors would like to thanks their respective families for endless support during the preparation of the book. In particular, Arman would like to express his thanks to his wife, Zhila, his parent, Parvin, and his brothers, Meharn and Houman for constant support and encouragement. Manoj would like to express his appreciation to his wife Sunanda, their son, Aniruddh, and daughter Arushi for their support.

                                        Arman Vassighi and Manoj Sachdev

Foreword

Semiconductor industry has benefited from several decades of growth following the Moore's law. Today's high-performance Integrated Circuits (ICs) have more than one billion transistors. Scaling of CMOS technology has enabled this increase in transistor count, and scaling continues in spite of emerging barriers in process technology development, design and test. Today, the 65nm CMOS technology node is moving from development to high volume manufacturing while research and development continues on future technology nodes.

This growth comes with multiple challenges. Design of ICs in these scaled technologies faces formidable limitations. It is becoming increasingly difficult to sustain supply and threshold voltage scaling in order to provide the required performance increase, limit energy consumption, control power dissipation, and maintain the reliability. These requirements pose several difficulties across a range of disciplines spanning process technology, manufacturing, circuits, testing, systems, and architecture. One of these critical challenges is the thermal and power management of high speed ICs which has been subject of extensive research over past several years. Moreover, reliability screening and burn-in at elevated voltage and temperature further exacerbate this problem. Furthermore, researchers have looked into finding ways to cool ICs efficiently by accounting for the cooling cost in total system power. It may be possible to use efficient cooling to improve performance, leakage, and reliability of ICs, and thus continue on the path of scaling.

This book is very timely to address these challenges and to capture key learning concepts. This book discusses thermal design barriers and the proverbial power wall that IC and system designers are facing today. We have followed authors' work, interacted with them, and we highly recommend this book to students, engineers, professionals and people that are pursuing advanced designs and are facing some of these challenges.

Vivek De

Chapter 1

INTRODUCTION

Abstract: The drive for higher performance has led to grater integration and higher clock frequency of integrated circuit chips in general and microprocessor chips in particular. This translates to higher power consumption and consequently higher heat dissipation and higher junction temperature. This chapter discusses the *CMOS* technology scaling concepts, power and thermal trend and junction temperature trends. Then the thermal issues will be introduced. Finally the motivation behind the preparation of this book will be explained.

Key words: *CMOS* Scaling, Power Trends, Junction Temperature Trends.

1.   EVOLUTION OF *CMOS* TECHNOLOGY

The basic concept of field effect transistor (*FET*) independently introduced by Lilienfeld, and Heil, respectively in 1930s [1, 2]. However, it took thirty more years to make the idea a reality, when in 1960 Kahng and Atala put the idea into practice in *Si-SiO$_2$*. Since then, Metal Oxide Semiconductor FET (*MOSFET*) has been the technology of choice for a vast majority of applications owing to its simplicity, inexpensive manufacturing process, integration capability, and extremely small power consumption compared to other integrated circuit technologies. In addition, the ability to improve performance with reduced power consumption per logic gate made Complementary *MOS* (*CMOS*) the dominant technology for integrated circuits.

1.1   Concept of Scaling

Transistor exponential scaling behavior which has come to be known as Moore's Law [3], is the primary factor driving speed and performance improvement in both microprocessors and memories [3-6]. Historically, *CMOS* technology scaling per technology node has:

- Reduced the gate delay by 30% allowing an increase in maximum clock frequency of 43%.
- Doubled the device density.
- Reduced the parasitic capacitance by 30%.
- Reduced the energy and active power per transition by 65% and 50%, respectively [7-9].

Figure 1.1 shows the evolution of Intel microprocessor operating clock frequency and gate delays per clock since 1987. To achieve this, transistor width, length, and oxide dimensions were scaled down by 30%. As a result, the chip area was decreased by 50% for the same number of transistors, and total parasitic capacitance was decreased by 30%. Figure 1.2 illustrates the concept of scaling. As it can be seen in this figure, all dimensions of a *MOS* transistor are scaled by a factor s ($s > 1$) to produce the next generation transistor with the same electrical behavior.

With scaling all the dimensions and voltages are reduced by a factor s, and doping densities are increased by s, therefore the electric field inside the device remains the same as before. Since the electric field in this kind of scaling is constant, it is known as *constant electric field scaling* (*CFS*). Constant electric field scaling prevents the device damage from excessive electric field. In the early generations of the *MOSFETs*, voltage was kept constant with scaling. Since the oxide thickness was scaling down, the electric field was increasing with scaling resulting in higher performance. Therefore, earlier devices (until 0.8 μm) followed the *constant voltage scaling* (*CVS*) path. However, it was subsequently abandoned in favor of *CFS* owing to higher electric fields inside the device and its implications on long-term device reliability.

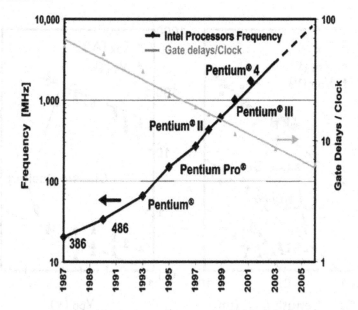

Figure 1-1. Processor frequency trend adopted from [8].

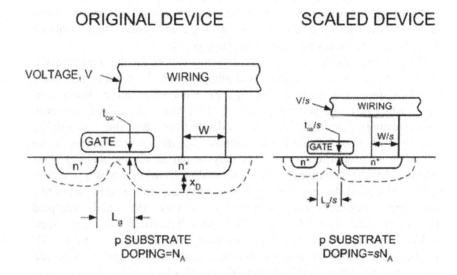

Figure 1-2. Concept of *MOS* Scaling.

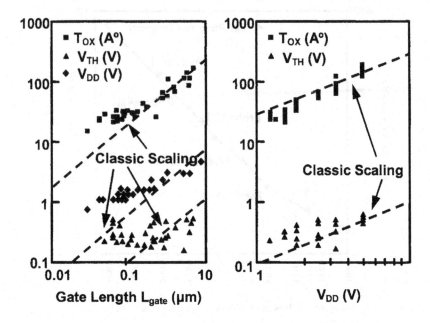

Figure 1-3. Published industry trends are compared to classic scaling [10].

Although, classic scaling has not been strictly followed by industry, it has been a blueprint for the major features over the period from roughly 1981 to 2001. Figure 1.3 shows published industry results for transistor gate-oxide thickness (T_{OX}), threshold voltage (V_{TH}), and power supply voltage (V_{DD}), all plotted versus the gate length (L_{GATE}) [10].

Dashed lines show the classic scaling trajectories for these parameters. Taking gate length as a measure of the lithography scale, it can be noted that V_{DD}, V_{TH}, and, to a lesser extent, T_{OX} have decreased more slowly than L_{GATE}, while I_{DSAT} has actually increased rather than remaining fixed (as in classic scaling). The figure on right-hand side shows the same V_{TH} and T_{OX} data as the figure on left-hand side. Note that T_{OX} and V_{TH} are relatively close to scaling in proportion to V_{DD} (as they would in classic scaling). This indicates that the deviations from classic scaling have been driven primarily by V_{DD}, which has decreased more slowly than L_{GATE}. In the early part of this time span (1μm to 0.5μm), a reluctance to leave the widely accepted industry-standard V_{DD}=5.0V, inherited from Transistor-Transistor Logic (*TTL*), substantially retarded V_{DD} reduction. As the transition to a 3.3V standard gained momentum, an increased emphasis on performance and power gave the circuit board designers more flexibility for scaling V_{DD} and also allowed *CMOS* process technology developers the freedom to optimize V_{DD} scaling for power and performance to a greater degree [10].

Table 1.1 summarizes some of the parameters which are changing with scaling in three different scaling scenarios. In this table General Scaling Scenario (*GSS*) refers to more recent scaling strategy that as mentioned above has been followed by industry, where voltage has scaled less aggressively by a factor g (where $s > g > 1$) because of non-scaling behavior of sub-threshold voltage and *off* current. *GSS* offers the performance benefits of *CFS* or *CVS* while its power dissipation is in between *CFS* and *CVS*.

Another observation from Table 1.1 is that the power density (s^2/g^2) is increasing with scaling. This is due to the fact that V_{DD} and I_{DSAT} are scaling with slower rate ($1/g^2$) than area ($1/s^2$). It must be noted that while the device area is reduced by scaling, the die area is increasing to accommodate higher number of transistors and consequently more complex circuits.

Table 1-1. Scaling concepts for *MOS* transistor.

Parameter	Relation	Constant Electric Field Scaling	Constant Voltage Scaling	General Selective Scaling
W, L, t_{ox}		$1/s$	$1/s$	$1/s$
V_{DD}, v_t		$1/s$	1	$1/g$
Area	WL	$1/s^2$	$1/s^2$	$1/s^2$
C_{ox}	$1/t_{ox}$	s	s	s
C_{gate}	WLC_{ox}	$1/s$	$1/s$	$1/s$
I_{sat}	$C_{ox}VW$	$1/s$	1	$1/g$
Gate Delay	VC_{ox}/I_{sat}	$1/s$	$1/s$	$1/s$
Power Dissipation	$I_{sat}V$	$1/s^2$	1	$1/g^2$
Power Density	Power/Area	1	s^2	s^2/g^2

Figure 1.4 shows the trend for the number of transistors in Intel microprocessors. It is expected that around year 2008-2010 the number of transistors in high performance microprocessors exceeds one billion [11]. This will help designers to add more functionality to the chips.

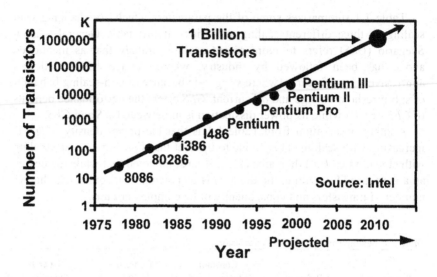

Figure 1-4. Number of transistors in Intel microprocessors [11].

1.2   Scaling and Power

The increase in number of transistors as mentioned before is due to the reduced transistor size and increased die size with scaling. Figure 1.5 shows the trend in die size from 1970 to 2010 [11]. Since the die size increase 7% per year and number of transistors per area are doubled in each new generation.

Unlike the transistor and die size trends, the total power of the microprocessors does not follow any scaling scenario. Figure 1.6 shows the power trends and power projections of Intel microprocessors from 1971 to 2008 [11]. In this figure the dotted line shows the power trend for classic scaling. It is evident from this figure that the power trend of the high performance microprocessors more or less has followed the classic scaling until the late 90s. After that the power is increasing exponentially in logarithmic scale. The main reason behind this is the increase in the sub-threshold leakage in recent years which is not included in total power projections in any of scaling scenarios.

The scaling of the threshold voltage is the primary cause for increasing leakage current. A linear reduction of approximately 80-100 mV in the threshold voltage increases the sub-threshold leakage by one order of magnitude. Over several technology nodes, the threshold voltage has been reduced by 500 mV or more resulting in five orders of magnitude increase in the transistor leakage current!! Furthermore, growing number of transistors is also adding to the overall chip level power consumption.

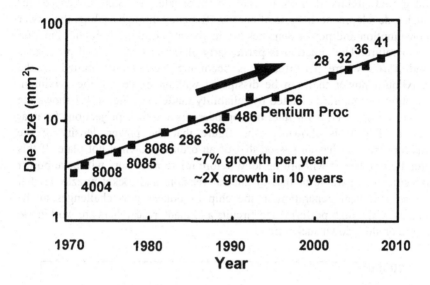

Figure 1-5. Die size trend for Intel microprocessors [11].

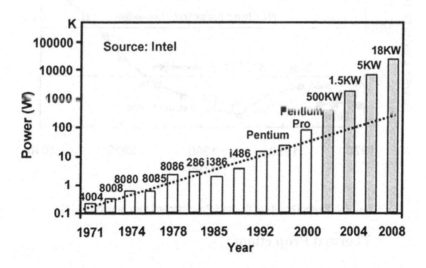

Figure 1-6. Power trends for Intel microprocessors [11].

2.   EMERGENCE OF THERMAL ISSUES

Thermal issues are becoming increasingly important for wide variety of integrated circuits. It is not difficult to appreciate the underlying causes for such a development. In spite of energy efficiency through scaling, the power consumption and power densities are increasing resulting in higher junction temperature. The situation is particularly alarming for high-end processors and *ASICs* where performance is becoming increasingly limited by the maximum power that can be dissipated without exceeding the maximum junction temperature dictated by reliability guidelines. Figure 1.7 shows the power density trends for Intel microprocessors and their projections for near future [11]. It is shown in the figure that the power density of the microprocessors has surpassed that of an ordinary kitchen hot plate. If this trend continues, it will not be long before microprocessors will have power densities comparable to that of nuclear reactors and rocket nozzle. Higher power and heat generation in the chip introduces new challenges to the design of the high performance circuits and many researchers are working to avoid or slow down such a trend.

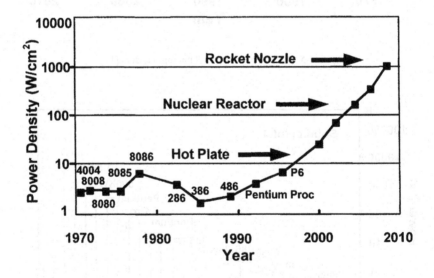

Figure 1-7. Power Density of the Intel Processor and their projection [11].

2.1   Thermal Projections

Most of failures mechanisms in semiconductor devices are temperature dependent processes meaning that higher junction temperature accelerate the thermally driven failure mechanisms. According to the International Technology Roadmap for Semiconductors (*ITRS*), for future generation of technologies, the junction temperature of the semiconductor devices must be kept at 85°C or lower to ensure the long term reliability of devices [12].

Table 1-2. ITRS sort and long term power and thermal projection for years 2003 to 2016 [12].

Year of Production	Near Term				Long Term			
	2003	2004	2005	2006	2007	2010	2013	2016
Maximum Junction Temperature (°C)								
Cost Performance	85	85	85	85	85	85	85	85
High Performance	85	85	85	85	85	85	85	85
Ambient Temperature (°C)								
Cost Performance	45	45	45	45	45	45	45	45
High Performance	45	45	45	45	45	45	45	45
Power (W)								
Cost Performance	81	85	92	98	104	120	138	158
High Performance	150	160	170	180	190	218	251	288
Required Thermal Resistance (°C/W)								
Cost Performance	0.49	0.47	0.43	0.41	0.38	0.33	0.29	0.25
High Performance	0.27	0.25	0.24	0.22	0.21	0.18	0.16	0.14

Table 1.2 lists the projected values for the maximum junction temperature, ambient temperature, and the power consumption (heat dissipation) in the near term and long term for cost performance and high performance market segments. One can compute the required junction to ambient thermal resistance as the ratio of the difference between junction and ambient temperatures and the chip power. Computed thermal resistance values are listed in the last two rows of the Table 1.2. These resistance values represent the required overall junction to ambient thermal resistance. For a given heat dissipation rate, a lower chip to ambient thermal resistance

ensures a lower package operating temperature and therefore a longer failure free operating period [13].

2.2   Thermal Issues

Junction temperature is one of the most important *CMOS* parameters, which impacts the performance, power and the reliability of the integrated circuits. Junction temperature during nominal operating and stress conditions affects the performance, the leakage power, and the long term reliability of the chip.

The performance of the integrated circuits is proportional to the driving current of the *CMOS* transistors. The driving current of the *CMOS* transistors is a function of the carriers' mobility. Increasing junction temperature decreases the carrier mobility, driving current of the *CMOS* transistor, and consequently degrades the performance of the circuit.

The leakage power, consisting of several components, is a strong function of the junction temperature. In higher junction temperature the leakage power increases and results in higher power consumption. The elevated leakage power in turn increases the junction temperature due to the extra power dissipation. In extreme cases this positive feedback between the junction temperature and the leakage power may lead the chip to thermal runaway where the chip will be destroyed due to excessive heat dissipation. The chip is more susceptible to the thermal runaway during the burn-in reliability screening test where the chips are tested under elevated supply voltage and junction temperature. Integrated circuit designers try to limit the leakage current by incorporating different techniques at circuit, architecture, and system levels.

The long term reliability is one of the key aspects of survival in this very competitive environment. Most of the failure mechanisms in *CMOS* integrated circuits are thermally activated processes. Higher junction temperature accelerates these failure mechanisms and reduces the lifespan of the chip. These failure mechanisms include, gate oxide breakdown, electro-migration, hot electron effects, negative bias temperature instability, etc. and are described in subsequent chapters.

3.   MOTIVATION OF THIS BOOK

Thermal issues have become a show stopper in our ability to further integrate and operate transistors. In other words, if these issues are not addressed properly, they will limit the scaling of technology into nano-metric regime. In this context, the primary motivation of this book is to make readers aware of thermal issues. Proper thermal management is the key to achieve high quality, reliability, and performance circuits. Reliability of integrated circuits depends exponentially on the junction temperature (temperature of the silicon). Even small differences in the junction temperature (of the order of 10-15°C) can result in a factor of 2x reduction in the device lifetime. The localized or distributed junction temperature may become extremely high and may lead to thermal runaway if proper thermal management is not done. Furthermore, thermal analysis is also important owing to cross-chip temperature gradients and thermal coupling effects induced by localized power dissipation which may affect the performance of the circuit. Performance degradation caused by thermally induced device mismatch is a major concern in the design of high speed and/or high precision integrated circuits such as *ALUs,* data converters, instrumentation amplifiers, analog multipliers, etc. Given the above, it is not surprising that the awareness of thermal issues and the need for thermal-design has increased over the past few years.

A great deal of attention has been paid to thermal issues recently; however, this knowledge is distributed in various conference and journal publications. Hence, it does not provide an overview to engineers and managers who are suddenly confronted with these issues. Moreover, lack of books in this area has not helped the cause of spreading the thermal awareness issues. This book makes an attempt to fill the void and provides a comprehensive overview of the power and thermal management.

In this book power and thermal management issues in integrated circuits during normal operating conditions and stress operating conditions are addressed and the latest research that has been carried to solve these problems is discussed. The book also presents the research in the area of electro-thermal modeling of integrated circuits. The electro-thermo models and associated *CAD* tools are presented and various techniques at the circuit and system levels are reviewed. The book also provides an insight into the concept of thermal runaway and how it may best be avoided. Finally the low temperature operation of integrated circuits is presented. This book will benefit researchers in the industry and academia in a wide area of VLSI design by reviewing the state of the art research in the area of power and thermal management of integrated circuits.

References

1. J. E. Lilenfeld, "Method and apparatus for controlling electric currents", *U.S. Patent no. 1,745,175*, 1926.
2. O. Heil, "Improvements in or relating to electrical amplifiers and other control arrangements and devices", British Patent no. 439,457, 1935.
3. G. Moore, "Cramming More Components into Integrated Circuits," *Electronics*, Vol. 38, No. 8, 1965.
4. R. H. Dennard, F.H. Gaensslen, H.N. Yu, V.L. Rideout, E. Bassous, and A.R. Leblanc, "Design of ion-implanted MOSFETs with very small physical dimensions," *IEEE Journal of Solid State Circuits*, vol. SC-9, pp. 256–268, Oct. 1974.
5. G. Baccarani, M.R. Wordeman, and R.H. Dennard, "Generalized Scaling Theory and its Application to a ¼ micrometer MOSFET design," *IEEE Transactions on Electron Devices*, vol. ED-31, pp. 452–462, Apr. 1984.
6. D.J. Frank, R.H. Dennard, E. Nowak, P.M. Solomon, Y. Taur, and H.P. Wong, "Device Scaling Limits of Si MOSFET and Their Application Dependencies," *Proceedings of the IEEE*, vol. 89, no. 3, Mar. 2001.
7. S. Borkar, "Design challenges of technology scaling", IEEE Microelectronics, pages 23-29, July-August 1999.
8. S. Rusu, "Trends and challenges in VLSI technology scaling toward100 nm".   ESSCIRC,   2001.web-page:http://www.esscirc.org/esscirc2001/C01-Presentati.ns/404.pdf.
9. S. Thompson, P. Packan, and M. Bohr, "MOS scaling: transistor challenges for the 21st century", Intel Technology Journal, Q3, pages 1-19, 1998. http://developer.intel.com/technology/itj/archive.htm.
10. E. J. Nowak. \Maintaining the bene¯ts of CMOS scaling when scaling bogs down". IBM Journal of Research and Development, Vol. 48, No. 2/3, 2002.
11. J. M. Rabaey, A. Chandrakasan, B. Nikolic, "Digital integrated circuits", Prentice Hall, 2003.
http://bwrc.eecs.berkeley.edu/Classes/IcBook/instructors.html
12. International Technology Roadmap for Semiconductors, 2001.
13. S. P. Gurrum, S. K. Suman, Y. K. Joshi, and A. G. Fedorov, " Thermal issues in next generation integrated circuits", IEEE Transaction on Device and Materials Reliability, Vol. 4, No. 4, pages 709-714, 2004.

Chapter 2

POWER, JUNCTION TEMPERATURE, AND RELIABILITY

Abstract: In this chapter, we initially discuss the power consumption trends and its impact on junction temperature elevation. Junction temperature plays a pivotal role in determining long term integrated circuit reliability. Therefore, we discuss the impact of junction temperature on the reliability of *CMOS* integrated circuits.

Key words: *CMOS*, Junction Temperature, Reliability, Leakage Power, Failure Mechanisms.

1.   POWER IN NANOMETER REGIME

The power of an integrated circuit (*IC*), for a fixed operating voltage and temperature, increases linearly with the clock frequency (the frequency of a master signal with which all operations must be synchronized) driving the *IC*. Extrapolation of the power vs. frequency response down to a frequency of zero (which may be realized in a sleep mode) yields a non-zero power, which is referred to as the static power, P_{static}. Increasing amount of attention is paid on to the techniques to reduce the overall power consumption, and an interested reader is referred to several outstanding work in this domain [1]. However, we briefly discuss various dynamic and leakage current components in subsequent sub-sections. This discussion will be useful in building intuitive understanding for following sections in this chapter.

1.1   Dynamic Power

The component of power which is proportional to the frequency is referred to as the dynamic power, $P_{dynamic}$. The dynamic power is due primarily to the charging and discharging of capacitances in the *IC*, and can be represented by an effective switching capacitance, C, via the well known relationship,

$$P_{dynamic} = C \cdot V_{DD}^2 \cdot f \qquad (2.1)$$

In this equation C does not necessarily represent the actual total capacitance being switched by the chip since many of the circuits may be switching at some fraction of f (or, for that matter, at some multiple of f). Furthermore, another source of active power, sometimes referred to as short-circuit; shoot-through, or crossover power, is also lumped into C. This short-circuit power is due to current which completes a path from the power supply node to ground directly through a network of n-type and p-type *FETs* during the short but finite time interval when the gates are close to $V_{DD}/2$, and hence both n- and p-type *FETs* are in a conducting state. Typically this component represents several percent of the active power.

1.2   Static Power

The standby current density increases exponentially as the channel length is decreased. This follows from the demand that V_{TH} decreases with V_{DD}, to maintain a high drive current and achieve the performance improvement. This current causes an additional power demand in the operation of *CMOS* which is often referred to as static power, since, unlike switching, or active power, static power is dissipated by all *CMOS* circuits all of the time, whether or not they are actively switching.

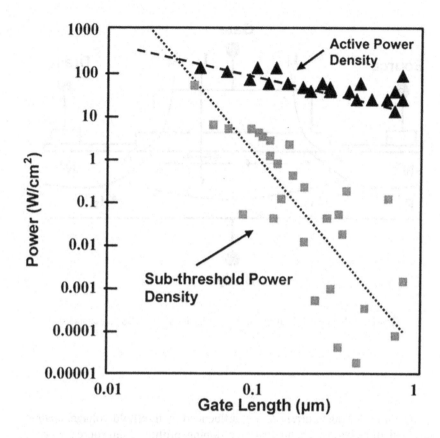

Figure 2-1. Static and dynamic power trends, vs. L_{gate} for a junction temperature of 25°C [2].

Figure 2.1 illustrates the static power trend based on sub-threshold currents calculated from the industry trends of V_{TH} , all for a junction temperature of $T_j = 25°C$. More practical values of T_j only serve to exacerbate this situation, with the off current of *MOSFETs* rising nearly two times for each *10°C* increase in T_j. For reference, the active power density is shown in Figure 2.1 in the same scale to illustrate that the sub-threshold component of power dissipation is emerging to compete with the long battled active power component for even the most power-tolerant, high speed *CMOS* applications. Empirical extrapolation (dashed curves) suggest that sub-threshold power will equal active power at L_{gate} of 20nm and this point will be encountered closer to L_{gate} of 50nm at elevated temperatures [2].

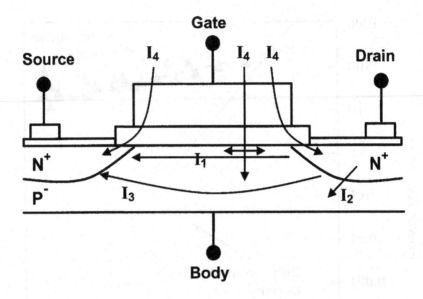

Figure 2-2. Figure 2.2: Leakage current mechanisms for *CMOS* transistors [3].

The total leakage current I_{OFF} is influenced by threshold voltage, channel physical dimensions, channel/surface doping profile, drain/source junction depth, gate oxide thickness, and V_{DD}. Understanding the different components of leakage current is a necessity for developing techniques to effectively reduce the off-state leakage. Figure 2.2 shows these leakage current mechanisms [3].

1.2.1   Subthreshold Leakage (I_1)

Subthreshold leakage is the weak inversion conduction current that flows between the source and the drain of a *MOS* transistor when gate voltage is below the threshold voltage, V_{TH}, [4]. Unlike the strong inversion region in which drift current dominates, the subthreshold conduction is dominated by the diffusion current. In a similar manner to charge transport across the base of a bipolar transistor, carriers move by diffusion along the surface. Weak inversion or subthreshold current typically dominates modern device off-state leakage due to low V_{TH}.

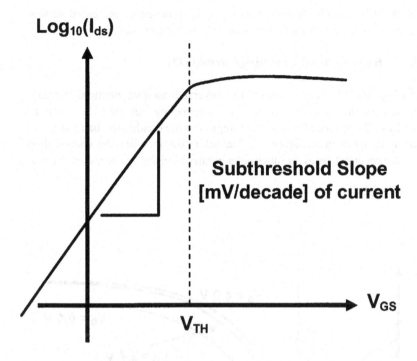

Figure 2-3. Subthreshold leakage in *NMOS* transistor.

Subthreshold current is exponentially related to the gate voltage as illustrated in Figure 2.3. The inverse of the slope of the $log_{10}(I_{ds})$ versus V_{gs} characteristic is called the subthreshold slope (S_t) [4]. Subthreshold slope indicates how effectively the transistor can be turned off when V_{gs} is reduced below V_{TH} thus it is desirable to reduce S_t. Typical values of S_t for bulk *CMOS* technology range from 70mV/decade to 120mV/decade (70 mV/decade means that 70 mV reduction in gate voltage reduces the subthreshold current by one order of magnitude).

Subthreshold current is an exponential function of the drain voltage. However, in long channel devices this dependency is very weak for very low V_{DS} (in the order of few $V_t = kT/q = 26mV$ at 25^0C). The subthreshold leakage increases with V_{DS} increase due to drain induced barrier lowering (*DIBL*) effect. When in short channel devices a high V_{DS} is applied to the drain, it lowers the potential barriers height. As a consequence more carriers can flow and give rise to higher subthreshold current. *DIBL* does not change the subthreshold slope (S_t), but it lowers the threshold voltage. Figure 2.4 shows that *DIBL* effect shifts the I_{DS}-V_{GS} curve to the left and to the top [3]. Devices with shorter channel length suffer more from *DIBL* effect. The effect of *DIBL* on subthreshold current can be reduced by increased surface and channel doping and shallower source/drain junction depths.

1.2.2   Band-to-Band Tunneling Current (I2)

If a high electric field is applied to a reverse biased *pn* junction, electrons tunnel across the junction giving rise to tunneling current. The electrons tunnel from the valance band of the *p*-region to the conduction band of the *n*-region as it is shown in Figure 2.5. Tunneling occurs when the voltage drop across the reverse biased *pn* junction is greater than the semiconductor band-gap.

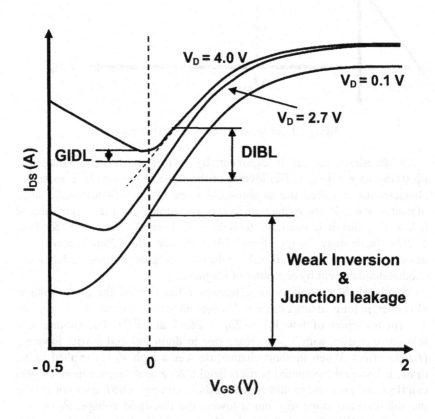

Figure 2-4. Figure 2.4: *n*-channel I_{DS} vs. V_{GS} showing *DIBL, GIDL,* weak inversion and *pn* junction leakage components [3].

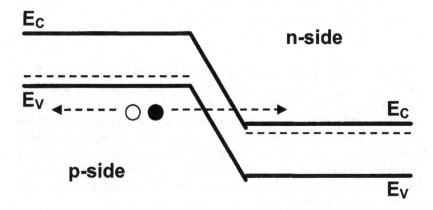

Figure 2-5. Band-to-Band tunneling in reversed bias *pn* junction[5].

When the drain (or source) of an *NMOS* device is biased at a higher voltage with respect to the substrate, *BTBT* current flows from the drain (or source) to substrate through the reverse biased *pn* junction [5]. In scaled *MOSFET* technologies, *p* and *n* regions are heavily doped, and halo implant is utilized to reduce the short channel effects (*SCE*). Owing to the heavily doped nature of junctions, the *BTBT* current increases significantly. The *BTBT* current can be a major component of the transistor off current. Reducing substrate doping near the substrate-drain/source junction is an effective method to reduce *BTBT* current, but it increases the short channel effects and leads to higher subthreshold leakage. An application of small forward body bias reduces the electric field across the junction, and hence reduces the *BTBT* current [6].

1.2.3   2.1.2.3 Punch through (I3)

As the channel becomes shorter in scaled *CMOS* technologies, the depletion layer widths of source and drain junction become comparable to the channel length. Using an abrupt one dimensional approximation, the width of the source junction W_S and that of the drain junction W_D are given as [7].

$$W_S = \sqrt{\frac{2\varepsilon_S}{qN_A}(V_{bi} + V_{BS})} \qquad (2.2)$$

$$W_D = \sqrt{\frac{2\varepsilon_S}{qN_A}(V_{DS} + V_{bi} + V_{BS})} \qquad (2.3)$$

Where ε_S is the permittivity of silicon, q is the electronic charge, N_A is doping of the drain and source region, V_{bi} is the built in potential and V_{DS} is drain to source voltage, and V_{BS} is the body to source voltage. If the transistor is off and if the V_{DS} is increased, at certain voltage the drain and source depletion layers will merge ($W_S+W_D = L$) resulting in the punch through. Under these conditions, the gate loses the control over the channel. Therefore, punch through is a major limitation of device operation in short channel *MOSFETs* and often retrograde implant is used to prevent the occurrence of the punch through.

1.2.4   Gate Oxide Tunneling (I4)

There are mainly two gate oxide leakage mechanisms between gate and the substrate of a *MOSFET*. One is the Fowler-Nordheim (*FN*) and another is the direct tunneling. Gate oxide leakage in scaled technologies is mainly due to the direct tunneling. In thin oxide layers (less than 3-4 nm), electrons tunnel directly from the silicon surface to the gate through the forbidden energy gap of the SiO_2 layer [5]. There are five different mechanisms that contribute to the direct tunneling current in *MOSFETs*. Two of these mechanisms cause the leakage between gate and source; and gate and drain extension overlap region, respectively. Next two mechanisms contribute leakage between the gate; and source and drain, respectively through the channel. Finally, the last component is the gate to substrate leakage current. The modeling of each of these components can be found in [8, 9].

2.   LEAKAGE REDUCTION TECHNIQUES

As illustrated in Figure 2.1, the leakage power is becoming significant component of the total power and may contribute to majority of the power dissipation in future *CMOS* technologies [10]. Therefore, in last decade and half a lot of attention has been paid on low power circuit and process techniques to reduce the elevated power consumption. Figure 2.6 shows the static or leakage power and dynamic power of the Intel microprocessors in different *CMOS* technologies. It is clear that the leakage power is increasing exponentially with the scaling.

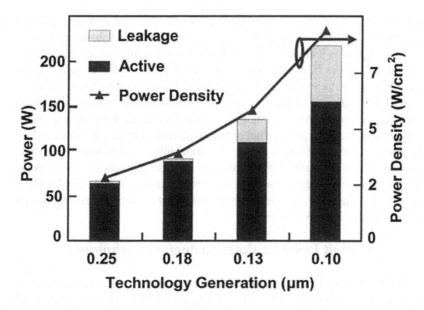

Figure 2-6. Static and dynamic power trends for Intel microprocessors [10].

Leakage current reduction can be achieved by utilizing either process techniques or circuit techniques. For low power applications combination of these two techniques are applied to reduce the leakage current. While at process level optimizing the device physical dimensions (length, oxide thickness, junction depth, etc) and the device doping profile lead to leakage current reduction, at circuit level, optimum biasing of the transistor terminals and other circuit techniques reduces the leakage current under different operating conditions. It must be noted that some of the leakage mechanisms like punch through can be optimized at process level while others can be optimized either at process level or at circuit level.

2.1 Process Level Leakage Reduction Techniques

Channel engineering is the process level technique which is used to reduce the leakage current while maximizing the linear and saturation drive currents. Changing the device doping profile in the channel region affects the electric field and potential contours and consequently changes the different current components. Steep retrograde wells and halo implants have been used to increase the drive current without affecting the off-state leakage current. Figure 2.7 shows the device with retrograde well and halo implants [6]. In a retrograde well structure, the doping concentration near the surface is low and it increases in subsurface region. This results in higher channel mobility on the surface and prevents punch through in the subsurface region.

Halo doping is a laterally non-uniform channel profile which controls the dependence of the threshold voltage on the channel length. In halo doping, the doping near the two edges of the channel is increased by the injection of point defects during sidewall oxidation, which gathers doping impurities from the substrate. As the channel length decreases, these highly doped regions consume a larger fraction of the total channel width, therefore reduce the width of the depletion region in the drain-substrate and source-substrate regions and guard against the threshold voltage degradation and leakage current increase. Although the halo implants reduce the subthreshold leakage current, they increase the band-to-band tunneling current in reverse bias *pn* junction due to increased doping.

2.2　　Circuit Level Leakage Reduction Techniques

Often process level leakage reduction techniques are not enough to achieve low leakage objectives in contemporary VLSIs. Hence, circuit level leakage reduction techniques are utilized. In general, circuit level techniques offer greater flexibility and can be optimized for specific applications. Moreover, these techniques can be utilized to reduce the leakage current of the circuit either under nominal or stress conditions. Stress conditions refer to reliability screening test environment, where *ICs* are subjected to higher junction temperature and higher supply voltage.

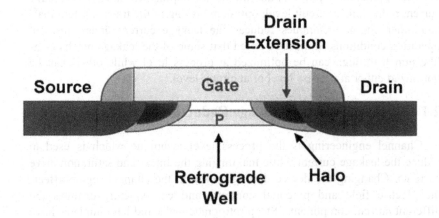

Figure 2-7. Different aspect of well engineering [6].

All the leakage reduction techniques under nominal conditions reduce the leakage current under nominal and stress conditions. There are some techniques that are designed specially for stress conditions, so under nominal conditions these techniques are disabled through proper signaling.

Major research has been carried out on low power and leakage current reduction [11-18]. The transistor I_{OFF} comprises of several different components [17, 3] of which the subthreshold current is the most dominant in scaled technologies. Subthreshold current of a *CMOS* transistor is modeled as follow:

$$I_{SUB} = A \cdot \exp\left(\frac{V_{GS} - V_{THo} - \gamma V_{SB} + \eta V_{DS}}{nV_t}\right) \cdot \left(1 - \exp\left(\frac{-V_{DS}}{V_t}\right)\right) \quad (2.4)$$

In this equation $A = \mu_o C_{ox} W/L_{eff} V_t^2 e^{1.8}$, μ_o is the zero bias carrier mobility, C_{ox} the gate oxide capacitance, L_{eff} is the transistor effective channel length, W is the transistor width, η is the drain induced barrier lowering coefficient, γ is the linearized body effect coefficient, n is the transistor subthreshold swing coefficient and V_t is the thermal voltage given by KT/q (~33mV at 110°C). In addition, V_{THo}, V_{GS}, V_{SB} and V_{DS} denote the transistor zero-bias threshold voltage, gate-source, source-body and drain-source voltages respectively. In the following sub-sections different techniques to reduce subthreshold current are described.

2.2.1   Non-Minimum L_{eff}

There is an exponential relationship between threshold voltage and effective channel length. Increasing the channel length will increase the threshold voltage. Figure 2.8 shows this exponential dependency. It must be noted that in very long channel lengths the dependency of threshold voltage on channel length decreases and extent of leakage reduction due to increased threshold voltage is limited by V_{TH} roll-off [14, 19].

In the region of interest, the threshold voltage increases almost linearly for small increases in the drawn channel length (L_{eff}). As a result, the increase in the transistor zero body bias threshold voltage is approximated as in Eq. (2.5) [19].

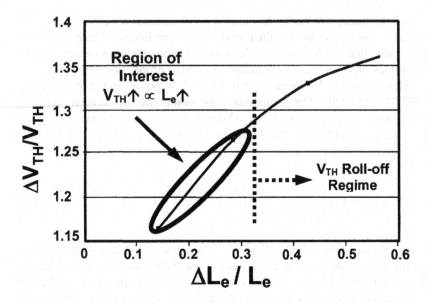

Figure 2-8. Normalized effective channel length vs. normalized threshold voltage.

$$\Delta V_{THo} = V_{THo}\left(\frac{\Delta L_{eff}}{L_{eff}}\right)$$ (2.5)

It must be noted that in nano-scale technologies, the channel mobility remains constant and independent of channel length due to velocity saturation. Therefore, the reduction in leakage current using non-minimum channel length transistors can be modeled as [19]:

$$\frac{\Delta I_{OFF}}{I_{OFF}} = 1 - \left[\frac{L_{eff}}{L_{eff} + \Delta L_{eff}} \cdot \exp\left(\frac{-\Delta V_{THo}}{nV_t}\right)\right]$$ (2.6)

Where, ΔL_{eff} is the change in effective channel length.

2.2.2   Stack Effect

Another solution to the increasing leakage is placing a non-stack transistor on a stack of two transistors without affecting the input load [20]. It has been shown that stacking two off-transistors significantly reduces the sub-threshold leakage compared to a single off-transistor. The drawback of this technique is the increased delay. This delay increase is comparable to high V_{TH} logic implementation in a dual V_{TH} technology. A significantly large fraction of the non-critical path implemented with this technique shows minimal performance degradation while reducing the sub-threshold leakage. The stack forcing technique can be either used in conjunction with dual V_{TH} or with a single V_{TH} technology [20].

Figure 2.9 shows a configuration of two stack transistors. The drain of the transistor *N1* is biased at V_{DD} and the source of the transistor *N2* is connected to the ground. The intermediate node voltage (V_N) reaches a steady state *DC* value for the two stack *N*-MOS pull down when both transistors are *OFF*. This value is within an order of magnitude of the thermal voltage (V_t). In 130 nm technology V_N is approximately 100mV and is due to $V_N = I_{OFF}R_{OFF}$ voltage drop across the bottom transistor (*N2*). As a result, a negative gate drive voltage (V_{GS}) appears across the top *N-MOS* transistor (*N1*) of the stack. Furthermore, there is a reduction in V_{DS} which suppresses the leakage current due to *DIBL* effect and also there is a negative body to source bias, V_{BS} (reverse body bias) which reduces the subthreshold current. Thus, the leakage current reduction using stack effect is as follow [19]:

$$\frac{\Delta I_{OFF}}{I_{OFF}} = 1 - \exp\left(\frac{-I_{OFF}.R_{OFF}.(1+\gamma+\eta)}{n.V_t}\right) \tag{2.7}$$

It must be noted that by using the stack transistor technique the drive current of the transistor *N1* is reduced accordingly. So it is important to place the stack transistors in the paths that charging the load is initial design target and discharging through the stack transistors with reduced drive current does not degrade the overall performance.

2.2.3 Reverse Body Bias (*RBB*)

The subthreshold leakage is reduced when the body of the transistor is biased to a negative voltage with respect to the source of the transistor ($V_{SB} < 0$). The reduction in the leakage current is proportional to the extent of the applied reverse body bias. However, beyond a certain optimal reverse body bias voltage, the transistor off-state leakage current starts to increase due to increased band-to-band tunneling current as shown in Figure 2.10 [15]. The optimum *RBB* for sub-130 nm technologies is approximately 30% of the V_{DD}.

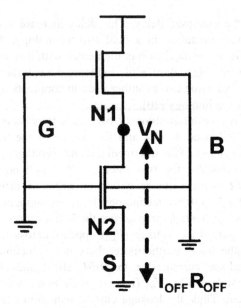

Figure 2-9. Leakage reduction using the stack effect.

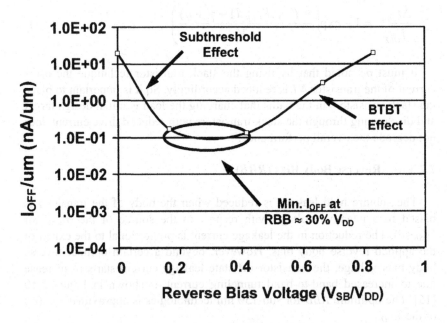

Figure 2-10. An optimum Reverse Body Bias (*RBB*) reduces the leakage current to its minimum [19].

For the region around the optimum reverse body bias, the leakage reduction can be modeled as [19]:

$$\frac{\Delta I_{OFF}}{I_{OFF}} = 1 - \exp\left(\frac{-\gamma . V_{SB}}{n.V_t}\right) \tag{2.8}$$

Reverse body bias technique is used to reduce leakage current during active operation, burn-in, as well as in standby mode. During active operation, *RBB* is applied to the idle portion of the chip to reduce overall chip leakage power without impacting the performance. Since the chip operational frequency is very low during burn-in, *RBB* can be applied to the whole chip simultaneously.

2.2.4 Multi-threshold Logic

This technique adjusts high performance critical path transistors with low V_{TH} while non-critical paths are implemented with high V_{TH} transistors. Hence, performance and power objectives are achieved at the cost of additional process complexity. Wei, et. al., reported a reduction of more than 80% in leakage power while meeting the performance objectives by using a dual V_{TH} technology [21].

Alternatively, a high V_{TH} transistor can be placed between power supply/ground and the high performance circuit or block (Figure 2.11). In the active mode, the high V_{TH} transistors are on and since their on-resistance is low, the performance impact is minimal. In the standby mode, the high V_{TH} transistor is off, and hence the leakage is limited to the leakage of a high V_{TH} transistor [22]. Traditionally, multi-threshold transistors are realized through different doses of threshold adjust ion implantations. Adjusting the threshold voltages can also be done by depositing two different oxide thicknesses or by different channel lengths [21].

2.2.5 Comparison of Leakage Reduction Techniques

Table 2.1 shows the leakage reduction achieved by using different techniques. In this table the theoretical models are compared with simulation results. Although theoretical models track simulation results for all techniques, they under estimate the reduction in leakage current. This is due to simplification in leakage reduction models. These results indicate that the stack effect technique reduces the leakage current by up to 12x while, the non-minimum L_{eff} technique (L_{eff} increased by 30%) reduces the leakage current by 9.3 x. Reverse body bias equal to 30% of V_{DD} reduces the leakage current by 2.2-2.3x [19].

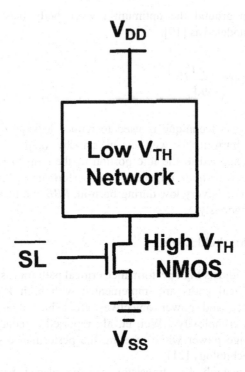

Figure 2-11. MTCMOS used to reduce the leakage of the low V_{TH} network.

Table 2-1. Leakage current reductions for 130nm technology [19].

Technique	Simulation Results	Theoretical Model
Non-minimum L_{eff} (L_{eff} +30%)	9.3x	8.7 x
Stack Effect	12.0 x	11.5 x
RBB (30% Reverse Bias)	2.3 x	2.1 x

2.2.6   Leakage Reduction and Impact on Performance

The leakage reduction is always desirable since it is wasted power. However, one must also look at the cost of reducing the power consumption with respect to its impact on performance. One of the most important trade-offs is leakage reduction and performance degradation trade-off. Design engineers often look at the I_{ON}/I_{OFF} ratio rather the absolute value of the I_{OFF} reduction when they examine any leakage reduction technique. The amount of I_{ON} determines the performance of the circuit and any reduction in I_{ON} leads to performance degradation.

Figure 2.12 shows the I_{ON}/I_{OFF} trends with technology scaling (solid lines) and also the threshold voltage trends with technology scaling (dash lines) for two cases of low and high threshold voltage. It can be seen that transistor $I_{OFF}/\mu m$ is increasing by 3-5x per generation resulting in the degradation of I_{ON}/I_{OFF} ratio with technology scaling. This results in excessive leakage currents for the 70 nm generation and offsets the reduction in switching energy obtained from scaling.

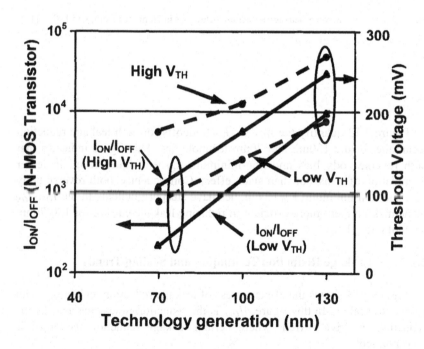

Figure 2-12. I_{ON}/I_{OFF} and V_{TH} scaling trends for typical corner device in 110°C and nominal V_{DD} [19].

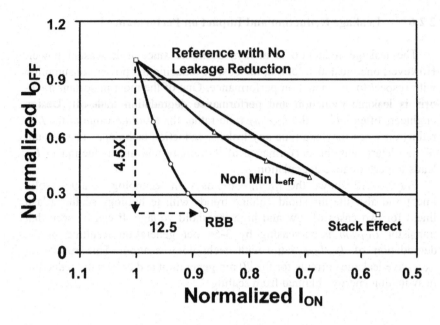

Figure 2-13. Comparison of leakage reduction techniques in 70 nm technology in 110°C [19].

Figure 2.13 quantify the I_{OFF} vs. I_{ON} tradeoffs for each leakage reduction technique for the 130nm and 70nm technologies. This figure indicates that both reverse body bias and non-minimum L_{eff} techniques result in lesser degradation of transistor than stack effect. Consequently, both reverse body bias and non-minimum L_{eff} techniques have steeper gradients in the I_{OFF}-I_{ON} plane making them more efficient in reducing leakage current for 130-70nm technologies [19].

2.2.7   Leakage Reduction Techniques and Scaling Trends

Figure 2.14 shows the effectiveness of leakage reduction techniques with respect to scaling. In this figure ΔI_{OFF} is the reduction in leakage and ΔI_{ON} is reduction in drive current where certain leakage reduction technique is incorporated.

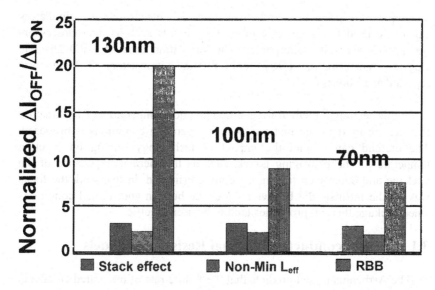

Figure 2-14. Trend for effectiveness of leakage reduction techniques with technology scaling.

It can be seen that for reverse body bias the ratio of $\Delta I_{OFF}/\Delta I_{ON}$ is the highest where, stack effect has the worse ratio. Another important observation from this figure is that the effectiveness of all these leakage reduction techniques is reducing with technology scaling. For reverse body bias technique, the $\Delta I_{OFF}/\Delta I_{ON}$ has decreased from 20x to 7.5x and for non-minimum L_{eff} it is reduced from 3.1x to 2.8x with scaling from 130 nm technology to 70 nm technology. This ratio for stack effect is reduced from 2.2x to 1.9x with scaling from 130 nm technology to 70 nm technology [19].

3. JUNCTION TEMPERATURE PROJECTIONS FOR DEEP SUB-MICRON TECHNOLOGIES

Several techniques can estimate junction temperature. One method directly measures junction temperature with thermal sensors at several on-chip locations during normal and burn-in conditions [23, 24]. Another method uses chip level 3D electro-thermal simulators that can find the steady-state *CMOS* VLSI chip temperature profile at the corresponding circuit performance [25, 26]. However, thermal sensors are relatively large devices, and accurate prediction requires a number of them placed on the *IC*. Sensors require calibration. Gerosa, et. al., reported a 0.2mm² thermal sensor with a sensing range of 0-128°C and a 5-bit resolution (4°C) [27]. Thermal sensors can only be used for verification, and one may have to use other techniques for prediction and estimation. 3D electro-thermal simulators cannot be used for large-scale integrated circuits such as microprocessors because of long simulation time. The simulation time of a 2D Discrete Cosine Transformation (*DCT*) chip (107,832 transistors, 8 MHz) was reported at 12 hours [26].

In this section, a method for average junction temperature (T_j) estimation that can be used for normal and burn-in operating conditions is proposed. The method can predict the impact of technology scaling on junction temperature. The packaging issues, such as the thermal impedance of the package and other such factors were not considered. In this work the focus was on the intrinsic die behavior under the burn-in and normal conditions since package thermal properties tend to be user-specific.

3.1   Semiconductor Thermal Resistance Models

The Arrhenius model predicts that the failure rate of integrated circuits is an inverse exponential function of the junction temperature. A small increase of 10-15°C in junction temperature may result in ~2x reduction in the life span of the device [28]. While T_a represents the ambient temperature for an *IC*, the relationship between ambient and average junction temperature for a VLSI is often described as in [29]:

$$T_j = T_a + P_{chip} \times R_{ja} \qquad (2.9)$$

Where T_a is the ambient temperature, P_{chip} is the total power dissipation of the chip, and R_{ja} is the junction-to-ambient thermal resistance. The impact of technology scaling on Eq. (2.9) must be analyzed to estimate the average junction temperature for several technologies. In this work the power dissipation and thermal resistance change with technology scaling were investigated in order to predict how these parameters will change.

The initial investigations on technology scaling and thermal resistance were carried out on bipolar transistors. For these devices, the thermal resistance was estimated as in [30]:

$$R_{ja} = \frac{1}{4K(L \times W)^{0.5}} \qquad (2.10)$$

Where K is the thermal conductivity of silicon, (LxW) is the emitter size, and R_{ja} is the thermal resistance (°C/mW). It was shown that the thermal resistance increased as the emitter size was reduced. Recently, a relationship between the thermal resistance of a *MOSFET* and its geometrical parameters was derived using a 3-D heat flow equation [31].

$$R_{ja} = \frac{1}{2\pi k}\left[\frac{1}{L}\ln\left(\frac{L+\left(W^2+L^2\right)^{0.5}}{-L+\left(W^2+L^2\right)^{0.5}}\right) + \frac{1}{W}\ln\left(\frac{W+\left(W^2+L^2\right)^{0.5}}{-W+\left(W^2+L^2\right)^{0.5}}\right)\right]$$

(2.11)

Where k is the thermal conductivity of silicon ($k = 1.5x10\text{-}4W/mm°C$ [32]), W and L are channel geometry parameters. The thermal conductivity of silicon has a temperature dependence described in [33].

The temperature dependence of silicon thermal conductivity is more important in silicon on insulator (*SOI*) technologies where self-heating contributes to a rise in junction temperature. So, our calculations assumed that the thermal resistance of silicon was temperature independent [31, 32]. Eq. (2.11) was used for the thermal resistance calculations for *MOSFETs* in different *CMOS* technologies.

3.2   Estimation of Junction Temperature Increase with Technology Scaling at Normal Conditions

F_{max} is defined as the maximum toggle frequency of an inverter in a given technology. The dynamic power consumption calculation under normal operating conditions was done at 70% of F_{max}. HSPICE simulations were carried out with BSIM model level 49. Transistor models for a 0.13μm *CMOS* technology were taken from United Microelectronics Corporation (*UMC*). Transistor models for other *CMOS* technologies were adapted from the Taiwan Semiconductor Manufacturing Corporation (*TSMC*). The simulation results and transistor sizes are given in Table 2.2.

The inverter load was the standard load element (*N-MOSFET*) used by *TSMC* for inverter ring-oscillator simulations. The load element sizes were taken from the *TSMC* and *UMC SPICE* models file specified for each of analyzed *CMOS* technologies. The International Technology Road map for Semiconductors (*ITRS*) 2002 [34] indicates that scaling down of device sizes is still in progress. Planar type transistors with 15-30 nm gate lengths have already been demonstrated [35]. However, 90-100 nm *CMOS* technology is currently the state-of-the-art for production of microprocessors and *SRAM* chips [36-38]. Therefore, we included the 90 nm *CMOS* technology node in our study of burn-in testing. The effective channel length of transistors for this technology was assumed to be 55-65 nm.

Table 2-2. Simulated *CMOS* inverter parameters and F_{max}.

CMOS Technology $\mu m/V_{DD}$, V	N-MOSFET W/L $\mu m / \mu m$	P-MOSFET W/L, Load $\mu m / \mu m$	N-MOSFET W/L, Load $\mu m / \mu m$	F_{max} MHz	$F_{operating}$ $=0.7F_{max}$ MHz
0.35/3.3	4/0.35	10.0/0.35	3.0/0.35	1450	1015
0.25/2.5	2.86/0.25	7.140.25	2.15/0.25	1950	1365
0.18/1.8	2.06/0.18	5.14/0.18	1.55/0.18	2300	1610
0.13/1.3	1.49/0.13	3.71/0.13	1.12/0.13	4000	2800

The total power consumption of an inverter toggling at $0.7F_{max}$ in four different technologies is simulated, with results given in Table 2.2. The thermal resistance of an average transistor was computed from Eq. (2.11). The average size of a transistor was estimated by averaging the *NMOS* and *PMOS* transistor widths.

As the transistor dimensions are reduced, the thermal resistance increases. Figure 2.15 illustrates inverter power dissipation at an operating frequency of $0.7F_{max}$ and the thermal resistance of an average transistor as functions of technology. Owing to lack of access to 90 nm *CMOS* technology, an alternative method was utilized to obtain the inverter power and thermal resistance estimates in Figure 2.15. For the 1.0 V, 90 nm *CMOS* technology, the *ITRS* predicts the transistor density in a microprocessor chip to be about 0.27 millions/mm^2. It is assumed that the transistor density is doubled with technology scaling for each new process generation. An industrial estimate of the power density of a microprocessor chip, implemented in 90 nm technology, is approximately 0.5W/mm^2 [37-39]. Power density is defined as the power dissipated by the chip per unit area under nominal frequency and normal operating conditions. Using these assumptions we can estimate the inverter power dissipation at normal operating conditions ($V_{DD} = 1$V, $T = 25$°C) and speed (Figure 2.15).

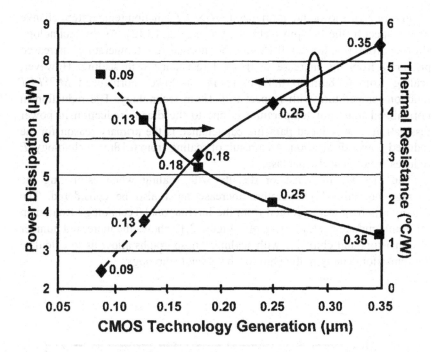

Figure 2-15. Inverter power dissipation and transistor thermal resistance for different *CMOS* technologies.

The scaling scenario of transistor sizes in a *CMOS* inverter was extended to 90 nm *CMOS* technology to calculate the thermal resistance. Transistor sizes of *PMOSFET (W/L)=3.0/0.1* and *NMOSFET (W/L)=1.0/0.1* were used. The calculated transistor thermal resistance for 90 nm technology using Eq. (2.5) is shown in Figure 2.15.

The 0.35μm *CMOS* technology was used as the reference technology. Using Eq. (2.9), ΔT can be defined as the temperature difference between junction and the ambient. If ΔT is set to unity for a 0.35μm technology, then the normalized change in ΔT with respect to the reference technology can be calculated. Using Eq. (2.9) and data presented in Figure 2.15, the normalized average temperature increase for different technologies was estimated. For example, Eq. (2.12) is used for calculation of $\Delta T_{0.25}=\Delta T_{0.35}$ ratio:

$$\frac{\Delta_{0.25}}{\Delta_{0.35}} = \frac{\left(T_j - T_a\right)_{0.25}}{\left(T_j - T_a\right)_{0.25}} = \frac{\left(P \times R_{ja}\right)_{0.25}}{\left(P \times R_{ja}\right)_{0.35}} \qquad (2.12)$$

Figure 2.16 shows the normalized *MOSFET* junction temperature change with respect to the 0.35µm technology using Eq. (2.12). As the technology shrinks from 0.35µm to 0.18µm, the normalized temperature increased primarily from the increase in thermal resistance with scaling. However, scaling from 0.18µm to 0.09µm results in lower normalized *MOSFET* junction temperature with respect to 0.18µm technology. The reduction in normalized transistor temperature is due to the drastic reduction in power dissipation. The reduced parasitic capacitance is the primary reason for the reduced power dissipation. As a result of scaling from 0.18µm technology, *P* reduces faster than R_{ja} increases.

The increase in transistor density with scaling when estimating the average normalized temperature increase must also be considered. The density numbers were adopted from the International Technology Road map for Semiconductors (*ITRS*) [34, 40]. Figure 2.17 shows the increased number of transistors and chip size with scaling. These graphs allow us to calculate the transistor density in the chip for the given technology.

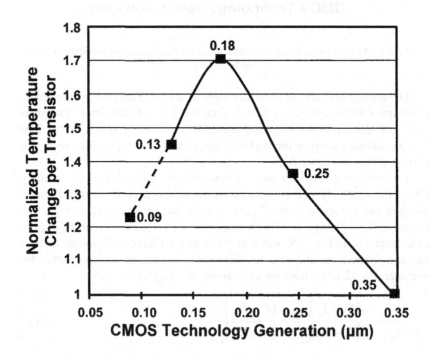

Figure 2-16. MOSFET junction temperature vs. technology.

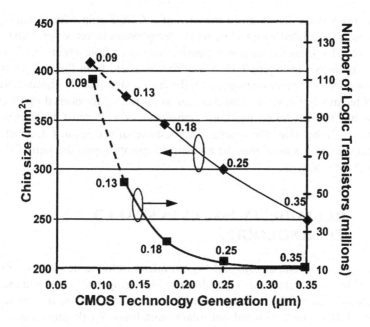

Figure 2-17. The trends of *CMOS* logic chips (data for graphs were adopted from [17, 20]).

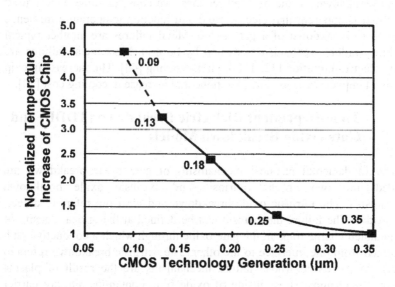

Figure 2-18. Normalized chip junction temperature increase with technology scaling for normal operating conditions.

The normalized temperature increase of a *CMOS* chip with technology scaling was calculated by multiplying the temperature increase per transistor in Figure 2.16 times the transistor density calculated from Figure 2.17. The results are shown in Figure 2.18. It can be concluded from Figure 2.18 that the normalized temperature increase of the chip is significantly elevated with *CMOS* technology scaling from 350 nm to 90 nm under normal operating conditions. The estimated junction temperature of a 90 nm *CMOS* chip is ~4.5 times higher than the junction temperature of a 0.35μm *CMOS* chip. This calculation assumed that the ambient temperature was the same for all analyzed technologies.

4.   RELIABILITY ISSUES IN SCALED TECHNOLOGIES

The effects of temperature and V_{DD} on microelectronic devices are often assessed by accelerated tests carried out at high temperature and voltage to generate reliability failures in a reasonable time period. Burn-in is often used as a reliability screen to weed out infant mortalities. Weak gate oxides are one of the major components of such failures. These failures are accelerated due to elevated electric field and temperature. Several dielectric breakdown models exist in the literature that can describe intrinsic as well as the defect-related breakdown. In the next subsections, we consider some widely used models. It is apparent that electric field and junction temperature influence the time to breakdown of a gate oxide. Metal failures are another typical reliability failure mechanism activated by burn-in. Most metal failures are due to electro-migration [42, 43] or stress voiding [43]. The increase in chip junction temperature results in an exponential increase in cooling cost [24].

4.1   Time-Dependent Dielectric Breakdown (TDDB) and Gate Oxide Breakdown Models

The fundamental physical mechanisms of gate oxide breakdown are divided into two groups: intrinsic and extrinsic oxide breakdown mechanisms. The intrinsic oxide breakdown and wear out refers to defect-free oxide. The failure mechanism can be defined at the critical density of accumulated charge traps in the gate oxide through which a conductive path is formed from one interface to the other. The extrinsic breakdown refers to defects in the oxide whose failure mechanisms are the result of plasma damage, mechanical stress inside of oxide film, contamination, hot carrier damage, or oxide damage by ion implantation. The extrinsic damages in gate oxide typically appear during relatively short time burn-in testing (~12 hours). Both breakdown mechanisms appear during burn-in as well as life testing [44, 45].

The *E* and *1/E* models are widely used in intrinsic gate oxide reliability predictions for oxide thickness greater than 50Å. Both models have a physical basis. The E-model is expressed as:

$$t = A \exp(-\gamma E) \exp\left(\frac{E_a}{kT_j}\right) \qquad (2.13)$$

Where *t* is the time to breakdown, *A* is a constant for a given technology, γ is the field acceleration parameter, *E* is the oxide field, E_a is the thermal activation energy, *k* is Boltzman's constant, and T_j is the junction temperature (°K). The *E*-model is based on thermo-chemical foundation and it indicates that increasing electric field across the gate oxide will decrease the time to break down. On the other hand, if we assume that the breakdown process is a current driven process, then the *1/E* model predicts:

$$t = \tau_0 \exp\left(\frac{G}{E}\right) \exp\left(\frac{E_a}{kT_j}\right) \qquad (2.14)$$

Where τ_0 and *G* are constants, *E* is the oxide electric field, E_a is the activation energy, and T_j is the junction temperature. The *1/E* model implies that the dielectric will not degrade in the absence of electric field. The *1/E* model ignores important thermal/diffusion processes that are known to degrade all materials over time, even in the absence of an electric field. Figure 2.19 shows the comparison of *E*-model and *1/E*-model failure in time (*FIT*) to time dependent dielectric breakdown (*TDDB*) in *T=175°C*.

To increase the drive current and to control the short channel effects, the oxide thickness should decrease at each technology node. The experimental measurements of time to breakdown of ultra thin gate oxides with thickness less than 40 Å show that the conventional *E* and *1/E TDDB* models cannot provide the necessary accuracy for calculation and prediction [46]. Hence, starting from about the 180 nm *CMOS* technology (T_{OX} range is about 26-31 Å) a new *TDDB* model is proposed [46, 47]. Experiments show that the generation rate of stress-induced leakage current (*SILC*) and charge to breakdown (Q_{BD}) in ultra- thin oxides is controlled by gate voltage rather than the electric field. This model (Eq. 2.15) includes the gate oxide thickness (T_{OX}) and the gate voltage (V_G) [48].

$$T_{bd} = T_0 \cdot \exp\left[\gamma\left(\alpha \cdot T_{OX} + \frac{E_a}{kT_j} - V_G\right)\right]$$

(2.15)

Where γ is the acceleration factor, E_a is the activation energy, α is the oxide thickness acceleration factor, T_0 is a constant for a given technology, and T_j is the average junction temperature. Time to breakdown physical parameter values were extracted from experiments as follows: $(\gamma \cdot \alpha) = 2.0$ $1/\text{Å}$, $\gamma = 12.5$ $1/V$ and $(\gamma \cdot E_a) = 575$ meV [48].

Historically, the activation energy has been an independent parameter in gate oxide breakdown models. However, starting from 130 nm technology, it becomes a function of accelerating electric field, as shown in Eq. (2.16) [49].

$$E_a \approx 1.15 - 0.07 \cdot E_{ox} \quad [eV]$$

(2.16)

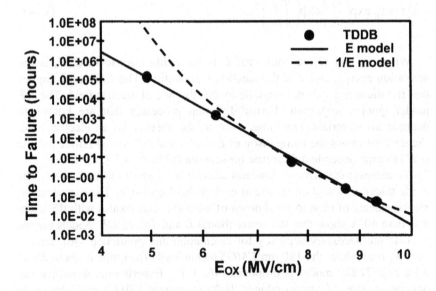

Figure 2-19. Comparison of *E*-model and *1/E*-model fit to T=175°C *TDDB*.

All the above methods describe the behavior of the intrinsic, good quality gate oxide. However, these models can also predict the time to breakdown under extrinsic oxide breakdown conditions, which include oxide damage by ion implantation, plasma damage, mechanical stresses, and contamination from technology processes. Under these conditions, the E_a is reduced. Since, the time to breakdown is a strong function of T_j and E_a, above mentioned oxide breakdown models can be used to predict the defect related breakdown.

To explain the time-dependent dielectric breakdown (*TDDB*) mechanism of extremely thin oxide films (~20-30 Å), researchers proposed two different approaches: (1) the anode hole injection model [50], (2) the electron trap generation model [51]. According to the first model, injected electrons generate holes at the anode that can tunnel back into the oxide. Intrinsic breakdown occurs when a critical hole concentration (Q_{BD}) is reached. The second model claims that a critical density of electron traps generated during stress is required to trigger oxide breakdown. Based on this model the breakdown event is presented as the formation of a conductive path of traps connecting the anode to the cathode interface. Recently, it was shown that the anode hole injection model and the electron trap generation model can be directly linked. A new model based on a percolation concept and statistical properties of oxide breakdown was developed [52]. Accordingly, breakdown can occur only when a connecting path of traps is formed across the gate oxide from the substrate to the gate due to the random defect generation throughout the insulating film. The physics-based analytical model [53], which is the extension and simplification of the common percolation concept, allows us to calculate (Eq. 2.17) the critical density (N_{crit}) of defects per unit of area at breakdown conditions as a function of gate oxide thickness (t_{ox}).

$$N_{crit}^{BD} = \frac{t_{ox}}{\alpha_0^3} \exp\left(-\frac{\alpha_0}{t_{ox}} \ln\left(\frac{A_{ox}}{\alpha_0^2} \right) \right) \tag{2.17}$$

Where, α_0 is the lattice constant of a cubic structure in the oxide bulk ($\alpha_0 \approx 2.34$ nm), and A_0 is the oxide area.

The relationship between the charge-to-breakdown (Q_{BD}), the critical defect density (N_{crit}), and the injected electron density (P_g) is [54]:

$$Q_{BD} = \frac{q N_{crit}^{BD}}{P_g} \tag{2.18}$$

The time-to-breakdown of thin oxides is determined by:

$$T_{BD} = \frac{Q_{BD}}{J_g} = \frac{qN_{crit}^{BD}}{P_g J_g} \qquad (2.19)$$

Where, J_g is the tunneling current across the gate oxide. The tunneling current (J_g) and the injected electron density (P_g) can be extracted from the experiments using *SILC* and *C-V* measurements [54]. Gate oxide defects have traditionally been a major reason for burn-in. Although other defects are activated during burn-in, it is important to understand the theory of oxide wear out and breakdown.

4.2   Electro-migration (EM)

Interconnect *EM* is the movement of metal atoms in the direction of electron flow due to momentum transfer from electrons to the metal ions under thermal and voltage stresses. *EM* is usually modeled by the empirical Black's formula [55], which relates the Mean-Time-To-Failure (*MTTF*) to the stressing conditions and is given as:

$$MTTF = A \cdot J^{-n} \exp\left(\frac{E_a}{kT_j}\right) \qquad (2.20)$$

Where A is the process constant dependent on material and geometry of the metal strip, n is a current exponent factor, T_j is the absolute junction (chip) temperature, k is the Boltzmann's constant, E_a is the activation energy and J is the current density. The activation energy for *Al-Cu* metal is in the range of 0.76-0.86 eV [56], and the activation energy for *Cu* interconnections, can vary widely from 0.7-0.9 eV to 1.0 eV. The lifetime of interconnects is decreased with the reduction of line width [57]. The accuracy of lifetime prediction is strongly dependent on the accuracy of the junction temperature measurement during the acceleration testing.

4.3   Hot Electron Effect

Substrate current has been successfully used as the Hot Carrier Injection (*HCI*) reliability indicator [58]. General substrate current model shows I_{sub} decreases while temperature increases. This is due to carrier mean free path reduction from more lattices scattering at higher temperature [59]. Recent study shows I_{sub} is insensitive to temperature over the range 77°K to 300°K due to the insensitivity of the carrier mean free path to the temperature. The latest study shows I_{sub} of 0.25µm device has reversed temperature dependence: I_{sub} increases with the temperature when V_{DD} is lower than a specific value, which can be defined as V_{DD} transition point. For *HCI* reliability, the field operation lifetime is projected from accelerated high bias stress data. The fully understanding of this reversed temperature effect and V_{DD} transition point will greatly impact on how to correctly set stress conditions and accurately project *HCI* lifetime to field operation conditions [60].

The widely used I_{sub} model is:

$$I_{sub} = I_d \frac{A_i \lambda}{\varphi_i} \left(V_{DD} - V_{Dsat} \right) . \exp\left(\frac{-\varphi_i}{\lambda k_m + \beta \frac{3}{2} kT} \right) \qquad (2.21)$$

Where I_d is the source drain current, A_i is the coefficient, λ is carriers mean free path, φ_i is the energy required to generate electron-hole pair. $\beta 3kT/2$ is the thermal kinetic energy which has been taken into account.

4.4　　Negative Bias Temperature Instability (NBTI)

Negative bias temperature instability (*NBTI*) is a *PMOS* degradation mechanism that can result in threshold voltage shifts up to 100 mV or more in very thin oxide devices [61][62]. Negative bias temperature instability can occur whenever a *PMOS* is biased in inversion. The damage has been shown to consist of both the formation of positive fixed charge and interface state generation [63][61]. The effects of these damage mechanisms are a negative V_{TH} shift (making the device harder to turn on) and transconductance degradation. Unlike hot carrier mechanisms, *NBTI* does not depend on any lateral (channel) field. The degradation is generally worst at $V_{DS} = 0$ [62][64]. Therefore, *NBTI* does not display significant channel length dependence, as hot carrier does.

Analog circuits and *SRAMs* are particularly sensitive to *NBTI* related problems. First of all, analog circuits often require matching among transistors, and utilize larger channel lengths to achieve good V_{TH} and I_D matching. Secondly, many of analog transistors, such as biasing, are constantly on making them prone to *NBTI*. *NBTI* could be expected to be the dominant degradation mechanism for these devices. Even for *PMOS* normally operating at a low gate bias (for which the *NBTI* shift would be small), there may be other circuit operation modes (such as power-down mode) which expose the devices to a relatively high V_{GS} [65].

5. SUMMARY

In this chapter active and leakage power consumption trends were reviewed. Subsequently, we discussed various strategies to reduce the total power consumption. The increased power consumption results in higher on chip, junction, temperature which in turns negatively influences various aging mechanisms. Therefore we discussed the impact of higher junction temperature on long term reliability.

References

1. D. M. Brooks, P. W. Cook, P. Bose, S. E. Schuster, H. Jacobson, P. N. Kudva, A. Buyuktosunoglu, J. Wellman, V. Zyuban and M. Gupta. "Power-aware microarchitecture: design and modeling challenges for next-generation microprocessors". IEEE Microelectronics, Vol. 20, No. 6, 2000.
2. E. J. Nowak. "Maintaining the benefits of CMOS scaling when scaling bogs down". IBM Journal of Research and Development, Vol. 48, No. 2/3, pages 26-44, 2002.
3. A. Keshavarzi, K. Roy and C.F. Hawkins. "Intrinsic leakage in deep submicron CMOS ICs: measurement based test solutions". IEEE Transactions on Very Large Scale Integration (VLSI) Systems, Vol. 8, No. 6, pages 717-723, 2000.
4. Y. Taur and T. H. Ning. "Fundamentals of modern VLSI Devices" Cambridge University Press, pages 120-128, 1998.
5. Y. Taur and T. H. Ning. "Fundamentals of modern VLSI Devices" Cambridge University Press, pages 94-95, 1998.
6. S. Tompson, P. Packan and M. Bohr. "MOS scaling: transistor challenges for 21st century". Intel Technology Journal, Q3, 1998.

7. S. M. Sze. "Physics of semiconductor device". John Wiley & Sons, 1936.
8. BSIM 4.2.1 MOSFET model, BSIM group, UC Berkeley, http://www-device.eecs.berkeley.edu/~bsim3/.
9. K. Cao, W. C. Lee, W. Liu, X. Jin, P. Su, S. Fung, J. An, B. Yu and C. Hu. "BSIM4 gate leakage model including source drain partition". Technical Digest of International Electron Devices Meeting, pages 815-818, 2000.

10. S. Rusu. "Trends and challenges in VLSI technology scaling toward 100 nm". Presented at ESSCIRC. [Online]. Available: http://www.esscirc.org/esscirc2001/C01_Presentations/404.pdf.

11. J. P. Halter, and F. N. Najm, "A gate-level leakage power reduction method for ultra-low power CMOS circuits". Proceedings of the IEEE CICC, pages 475-478, 1997.

12. T. Sakurai , H. Kawaguchi , T. Kuroda. "Low-power CMOS design through VTH control and low-swing circuits". Proceedings of the international symposium on Low power electronics and design, pages 1-6, 1997.

13. M. R. Stan. "Optimal Voltages and Sizing for Low Power". Proceedings of the 12th International Conference on VLSI Design - 'VLSI for the Information Appliance', pages 428, 1999.

14. N. Sirisantana, L. Wei and K. Roy. "High-Performance Low-Power CMOS Circuits Using Multiple Channel Length and Multiple Oxide Thickness". Proceedings of the 2000 IEEE International Conference on Computer Design: VLSI in Computers & Processors, pages 227-232, 2000.

15. A. Keshavarzi , S. Ma , S. Narendra , B. Bloechel , K. Mistry , T. Ghani , S. Borkar , V. De. "Effectiveness of reverse body bias for leakage control in scaled dual Vt CMOS ICs". Proceedings of the 2001 International Symposium on Low Power Electronics and Design, pages 207-212, 2001.

16. T. Kuroda, T. Fujita, S. Mita, T. Nagamatsu, S. Yoshioka, K. Suzuki, F. Sano, M. Norishima, M. Murato, M. Kako, M. Kinugawa, M. Kakumu, and T. Sakurai. "A 0.9V, 150-MHz, 10-mW, 4mm2, 2-D discrete cosine transform core processor with variable threshold-voltage (VT) scheme," IEEE Journal of Solid State Circuits, Vol. 31, No. 11, pages 1770-1779, 1996.

17. A. P. Chandrakasan , W. J. Bowhill , F. Fox. "Design of high-performance microprocessor circuits". Wiley-IEEE Press, 2000.

18. A. P. Chandrakasan and R. W. Brodersen. "Low power digital CMOS design". Kluwer Academic Publishers, Norwell, MA, 1995.

19. B. Chatterjee, M. Sachdev, S. Hsu, R. Krishnamurthy, S. Borkar. "Effectiveness and scaling trends of leakage control techniques for sub-130nm CMOS technologies". Proceedings of the International Symposium on Low Power Electronics and Design, pages 122-127, 2003.

20. S. Narendra, S. Borkar, V. De, D. Antoniadis, and A. Chandrakasan. "Scaling of stack effect and its application for leakage reduction". Proceedings of International Symposium of Low Power Electronics and Design (ISLPED), pages 195–200, 2001.

46 *Thermal and Power Management of Integrated Circuits*

21. L. Wei, Z. Chen, K. Roy, M. C. Johnson, Y. Ye, and V. K. De. "Design and optimization of dual-threshold circuits for low-voltage low-power applications". IEEE Transactions on VLSI Systems, Vol. 7, pages 16–24, Jan. 1999.
22. L. Wei, K. Roy, and V. K. De. "Low voltage low power CMOS design techniques for deep submicron ICs". Proceedings of International Conference of VLSI Design, pages 24–29, 2000.
23. T.J. Goh, A.N. Amir, C.P. Chiu, and J. Torresola. "Novel thermal validation metrology based on non-uniform power distribution for Pentium III Xeon cartridge processor design with integrated level two cache". Proceedings of Electronic Components and Technology Conference, pages 1181-1186, 2001.
24. S.H. Gunter, F. Binns, D.M. Carmean, and J.C. Hall. "Managing the impact of increasing microprocessor power consumption". Intel Tech. Journal, Q1:1-9, 2001.
http://developer.intel.com/technology/itj/archive.htm.
25. Y.K. Cheng, C.C. Teng, A. Dharchoudhury, E. Rosenbaum, and S.M. Kang. "A chip-level electrothermal simulator for temperature profile estimation of CMOS VLSI chips". Proceedings of International Symposium on Circuit and Systems, pages 580-583, 1996.
26. C.C. Teng, Y.K. Cheng, E. Rosenbaum, and S.M. Kang. "ITEM: A temperature dependent electromigration reliability diagnosis tool". IEEE Transaction on Computer Aided Design of Integrated Circuits and Systems, Vol. 16, No. 8, pages 882-893, 1997.
27. P. Reed, M. Alexander, J. Alvarez, M. Brauer, C.C. Chao, C. Croxton, L. Eisen, T. Le, T. Ngo, C. Nicoletta, H. Sanchez, S. Taylor, N. Vanderschaaf, and G. Gerosa. "A 250-MHz 5-W PowerPC microprocessor with on-chip L2 cash controller". IEEE Journal of Solid-State Circuits, Vol. 32, No. 11, pages 1635-1649, 1997.
28. R. Viswanath, V. Wakharkar, A. Watwe, and V. Lebonheur. "Thermal performance challenges from silicon to systems". Intel Technology Journal, Q3, pages 1-16, 2000.
http://developer.intel.com/technology/itj/archive.htm.
29. P. Tadayon. "Thermal challenges during microprocessor testing". Intel Technology Journal, Q3, pages 1-8, 2000.
30. R.C. Joy and E.S. Schlig. "Thermal properties of very fast transistors". IEEE Transaction on Electron Devices, ED-Vol. 17, No. 8, pages 586-594, 1970.
31. N. Rinaldi. "Thermal analysis of solid-state devices and circuits: an analytical approach". Solid-State Electronics, Vol. 44, No. 10, pages 1789-1798, 2000.

32. N. Rinaldi. "On the modeling of the transient thermal behavior of semiconductor devices". IEEE Transaction on Electron Devices, Vol. 48, No. 12, pages 2796-2802, 2001.

33. D.L. Blackburn and A.R. Hefner. "Thermal components models for electro-thermal network simulation". Proceedings of 9th IEEE SEMI-THERM Symposium, pages 88-98, 1993.

34. International Technology Roadmap for Semiconductors (ITRS). http://public.itrs.net/.

35. B. Doyle, R. Arghavani, D. Barlage, S. Datta, M. Doczy, J. Kavalieros, A. Murthy, and R. Chau. "Transistor element for 30 nm physical gate lengths and beyond". Intel Technology Journal, Vol. 6, No. 2, pages 42-54, 2002.

36. S.F. Huang, C.Y. Lin, Y.S. Huang, T. Schafbauer, M. Eller, Y.C. Cheng, S.M. Cheng, S. Sportouch, W. Jin, N. Rovedo, A. Grassmann, Y. Huang, J. Brighten, C.H. Liu, B.V. Ehrenwall, N. Chen, J. Chen, O.S. Park, and M. Common. "High-performance 50 nm CMOS devices for microprocessors and embedded processor core applications". IEDM, pages 237-240, 2001.

37. A. Ono, K. Fukasaku, T. Hirai, S. Koyama, M. Makabe, T. Matsuda, M. Takimoto, Y. Kunimune, N. Ikezawa, Y. Yamada, F. Koba, K. Imai, and N. Nakamura. "A 100 nm node CMOS technology for practical SOC application requirement". IEDM, pages 511-514, 2001.

38. D.J. Frank. "Power-constrained CMOS scaling limits". IBM Journal of Research and Development, Vol. 46, No. 2/3, pages 235-244, 2002.

39. S. Borkar. "Design challenges of technology scaling". IEEE Micro, pages 23-29, July-August 1999.

40. D.P. Valett and J.M. Soden. "Finding fault with deep-submicron ICs". IEEE Spectrum, pages 39-50, October 1997.

41. B. Davari, R.H. Dennard, and G.G. Shahidi. "CMOS scaling for high performance and low power - The next ten years". Proceedings of the IEEE, Vol. 83, No. 4, pages 595-606, 1995.

42. E. T. Ogawa, Ki-Don Lee, V. A. Blaschke, and P. S. Ho. "Electromigration Reliability Issues in Dual-Damascene Cu Interconnections". IEEE Transaction on reliability, Vol. 51, No. 4, pages 403-419, 2002.

43. C.F. Hawkins, A. Keshavarzi, and J.M. Soden. "Reliability, test and IDDQ measurements". IEEE International Workshop on Iddq testing, pages 96-102, 1997.

44. J.W. McPherson, V.K. Reddy, and H.C. Mogul. "Field-enhanced Si-Si bond-breakage mechanism for time-dependent dielectric break-down in thin-film SiO2 dielectrics". Applied Physics Letter, Vol. 71, No. 8, pages 1101-1103, 1997.

45. A.M. Yassine, H.E. Nariman, M. McBride, M. Uzer, and K.R. Olasupo. "Time dependent breakdown of ultra-thin gate oxide". IEEE Transaction on Electron Devices, Vol. 47, No. 7, pages 1416-1420, 2000.

46. J.H. Suehle. "Ultra thin gate oxide reliability: Physical models, statistics, and characterization". IEEE Transaction on Electron Devices, Vol. 49, No. 6, pages 958-971, 2002.

47. P.E. Nicollian, W.R. Hunter, and J.C. Hu. "Experimental evidence for voltage driven breakdown models in ultra thin gate oxides". Proceedings of IEEE International Reliability Physics Symposium, pages 7-15, 2000.

48. F. Monsieur, E. Vincent, D. Roy, S. Bruyere, G. Pananakakis, and G. Ghibaudo. "Time to breakdown and voltage to breakdown modeling for ultra-thin oxides (TOX < 32A°)". Proceedings of IEEE International Reliability Workshop (IRW), pages 20-25, 2001.

49. M. Kimura, "Field and temperature acceleration models for time-dependent dielectric breakdown". IEEE Transaction on Electron Devices, Vol. 46, No. 1, pages.220-229, 1999.

50. I.C. Chen, S. Holland, K.K. Young, C. Chang and C. Hu, "Substrate hole current and oxide breakdown". Applied Physics Letter, Vol. 49, No. 11, pages 669-671, 1986.

51. P.P. Apte and K.C. Saraswat, "Modeling ultra thin dielectric breakdown on correlation of charge trap-generation to charge-to-breakdown". Proceedings of International Reliability Physics Symposium, 1994, pages 136-142.

52. R. Degraeve, G. Groeseneken, R. Bellens, J.L. Ogier, M. Depas, P. J. Roussel, and H.E. Maes, "New insights in the relation between electron trap generation and the statistical properties of oxide breakdown". IEEE Transaction on Electron Devices, Vol. 45, No. 4, pages 904-911.

53. J. Sune, "New physics-based analytical approach to the thin-oxide breakdown statistics". IEEE Electron Device Letters, Vol. 22, No. 6, pages 296-298, 2001.

54. J.H. Stathis, "Physical and predictive models of ultra thin oxide reliability in CMOS devices and circuits". Proceedings of International Reliability Symposium, 2001, pages 132-149.

55. J. Black. "Electromigration: A brief survey and some recent results". IEEE Transaction on Electron Devices, ED-Vol. 16, No. 4, pages 338-347, 1969.

56. W.B. Loh, M.S. Tse, L. Chan, and K.F. Lo. "Wafer-level electromigration reliability test for deep submicron interconnect metallization". IEEE Hong Kong Electron Device Meeting, pages 157-160, 1998.

57. C.K. Hu, R. Rosengerg, H.S. Rathore, D.B. Nguyen, and B. Agarwala. "Scaling effect on electromigration in on-chip Cu wiring". IEEE International Interconnect Technology Conference, pages 267-269, 1999.

58. W. Wang, C.F. Hsu, L.P. Chiang, N.K. Zous, T.S. Chao and C.Y. Chang, "Voltage scaling and temperature effects on drain leakage current degradation in hot carrier stressed N-MOSFET" Proceedings of Reliability Physics Symposium, pages 209-212, 1997.

59. H. Sasaki, "Hot-carrier induced drain leakage current in N-channel MOSFET". IEEE Transaction on Applied Superconductivity, pages 726-729, 1987.

60. R. Bouchakour, L. Hardly, I. Limbourg and M. Jourdain, "Modeling and characterization of the MOSFET transistor stressed by hot-carrier injection". Proceedings of the 38th Midwest Symposium on Circuits and Systems, Vol. 1, Pages 61-64, 1995.

61. S. Ogawa, M. Shimaya, and N. Shiono, "Interface-trap generation at ultrathin SiO2 (4–6 nm)-Si interfaces during negative-bias temperature aging", Journal of Applied Physics, Vol. 77, No. 3, pages 1137–1148, 1995.

62. P. Chaparala, J. Shibley, and P. Lim, "Threshold voltage drift in PMOSFETs due to NBTI and HCI", IEEE International Integrated Reliability Workshop Final Report, pages 95-97, 2000.

63. C. E. Blat, E. H. Nicollian, and E. H. Poindexter, "Mechanism of negative-bias-temperature instability", Journal of Applied Physics, Vol. 69, No. 3, pages 1712–1720, 1991.

64. A. Acovic, E. Crabbé, F. Guarin, G. La Rosa, J. Lukaitis, and S. Rauch, "NBTI-channel hot carrier effects in pMOSFETs in advanced CMOS technologies", Proceedings of IEEE International Reliability Physics Symposium pages 282–286, 1997.

65. R. Thewes, R. Brederlow, C. Schündler, P. Wieczorek, A. Hesener, B. Ankele, P. Klein, S. Kessel, and W. Weber, "Device reliability in analog CMOS applications," IEEE Transaction on Electron Devices, Vol. 45, pages 2505–2513, 1998.

Chapter 3

BURN-IN AS A RELIABILITY SCREENING TEST

Abstract: Burn-in as a reliability screening test is one of the most important tests to ensure the quality and reliability of products under normal operating conditions. In this chapter subtle aspects of burn-in test are described and the junction temperature trends with scaling under burn-in condition are presented. Moreover, the packaging consideration and cooling solution for burn-in environment and optimization of burn-in conditions for maximum reliability and yield is reviewed.

Key words: Burn-in, Stress Testing, Acceleration Factor, Packaging, Cooling, Yield.

1. BURN-IN

The total power consumption of high performance microprocessors increases with scaling. Off state leakage current is an increasing percentage of the total current at the 130 nm and sub-100 nm nodes under nominal conditions. The ratio of leakage to active power becomes adverse under burn-in conditions and the off state leakage can become the dominant power. Typically, clock frequencies are kept in the tens of *MHz* range during burn-in resulting in a substantial reduction in active power. On the other hand, the voltage and temperature stresses cause the off state leakage to be the dominant power component.

Stressing during burn-in accelerates the defect mechanisms responsible for early life failures. Thermal and voltage stresses increase the junction temperature resulting in accelerated aging. Elevated junction temperature, in turn, causes leakage to further increase. In many situations, this may result in positive feedback leading to thermal runaway. Such situations are more likely to occur as technology is scaled down to the nano-meter regime. Thermal runaway increases the cost of burn-in dramatically. Other than thermal runaway, another issue with over stressing the chip is that the useful life of the chip will be shorter than it was planned for and this raises the long-term reliability issues. Hence, the temperature and voltage stress must be carefully optimized and tailored for any chip exposed to burn-in conditions.

2.   WHAT IS BURN-IN?

Component failure mechanisms and failure phenomena have been studied for a long time. Through experience and much data gathered by researchers and practitioners, component failure rates have been shown to follow the traditional bathtub curve.

The traditional bathtub curve (Figure 3.1) depicts component life in three stages. During the first stage, the failure rate begins high and decreases rapidly with time. This stage is known as the infant mortality period, and it has a decreasing failure rate (*DFR*). The infant mortality is mostly due to latent reliability defects. The infant mortality period is followed by a steady-state failure rate period, which is usually long and has a constant failure rate (*CFR*). This second stage is called the normal operating life and this is the period that the device will operate under normal conditions. Finally, the curve ends in the third stage, a period of wear out with an increasing failure rate (*IFR*). This is the period of aging. It is common for electronic devices to follow the traditional bathtub failure pattern.

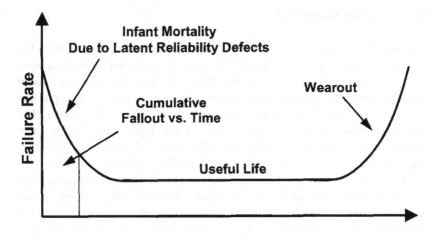

Figure 3-1. Bathtub curve.

2.1 Infant Mortality

Generally, a reasonable definition of the infant mortality period includes all failures prior to the normal operating period of the device life with its relatively stable and low steady state failure rate. The infant mortality period of the life cycle results from failures in a weak sub-population of the devices. The percentage of the weak sub-population (usually a small percentage), varies with the component type and the manufacturing lot, even for the same manufacturer. Factors contributing to the infant mortality include:

- Surface anomalies, for example: corrosion, contamination, and electromigration.
- Moisture entry.
- Quality defectives such as poor workmanship, irregularities, and process deviations.
- Electrostatic discharge.
- Random failures.

The above problems cannot be entirely eliminated, although good design and manufacturing help considerably. The distribution function of the infant mortality stage has been modeled as a Weibull distribution, a log-normal distribution, a non-homogeneous Poisson process, and a empirical distribution.

2.2   Why Burn-in?

In principle, burn-in is a process of eliminating defective parts from the production batch. The final tests that separate functional *ICs* from non-functional ones, in effect, are screening tests. However, *ICs* with defects that function marginally may not be eliminated by such screens and would end up in the field and begin to fail very early in the life of the system. The failure of these weak parts gives rise to the high initial failure rates commonly observed in the infant mortality period. A process of detection and elimination of such devices is called reliability screening.

Burn-in is a reliability screening method which requires acceleration of the mechanisms that give rise to infant mortality. The concept of the screening process is to accelerate the failures until the surviving population would begin its operational life with the low failure rates corresponding to the middle steady region of the bathtub curve. Temperature, voltage bias, and a combination of the two are often used as stresses to accelerate failures. The test conditions are selected depending on the nature and degree of the failure mechanisms causing infant mortality.

2.3   Burn-in Procedure

Traditionally, the burn-in procedure is executed prior to a final functional test procedure that weeds out the parts that have impaired functionality and/or high leakage current from the stresses during burn-in. Burn-in systems are designed to test hundreds of units in parallel over a period of many hours with operating frequencies in the tens of *MHz* range. There are three basic implementation methods for burn-in:

- Final package burn-in, where dies are packaged into their final destination packages and are subjected to burn-in at temperatures within the package thermal design constraints.
- Die level burn-in, where dies are placed into temporary carriers before they are actually packaged into their final form, thus reducing the cost of waste associated with added packaging.
- Wafer level burn-in (*WLBI*), where dies are tested while still in wafer form.

The last method potentially offers the greatest cost savings by eliminating the packaging waste cost. The first method offers the most reliable final product since package-related reliability issues are also taken into account. However, this method is expensive since fewer packaged devices can be burnt-in simultaneously, and post burn-in loss includes packaging cost. *WLBI* is relatively inexpensive, but it results in a relatively less reliable product since packaging related reliability issues are not addressed. Finally, the die-level burn-in with temporary carriers offers a compromise between the other two methods.

2.4 Static and Dynamic Burn-in

In static burn-in, dies are loaded into burn-in board (*BIB*) sockets; the *BIBs* are placed in the burn-in oven. The burn-in system applies power to the devices and heats them to 125°C-150°C for periods ranging from 12 to 24 hours. In static burn-in, the device under test (*DUT*) is powered but inputs are not toggled.

Dynamic burn-in mimics the static burn-in process, but also stimulates the *DUT* address, data, and clock inputs at a reduced rate (10-30 MHz) determined by the relatively cheap electronics of the burn-in tester. Under dynamic conditions, circuit nodes are toggled ensuring that voltage stress is applied to various transistors. Neither static nor dynamic burn-in monitors the *DUT* responses during the stress. Weak die destroyed by the burn-in process are not detected until a subsequent functional test stage. "Intelligent" burn-in systems not only apply power and signals to *DUTs*; they also monitor the *DUT* outputs. The Test During Burn-in (*TDBI*) method can guarantee that devices undergoing burn-in are indeed powered and that input test vectors are being applied. In addition, *TDBI* can perform some test functions. Detailed information about different burn-in methods and features of burn-in ovens can be found elsewhere [1-3].

3. TEMPERATURE AND VOLTAGE ACCELERATION FACTORS

Several industrial reliability standards are based on temperature and voltage acceleration factor models. The *MIL-HDBK-217F* US military standard defines the temperature acceleration factor as [4]:

$$\pi_T = 0.1 \cdot \exp\left(-A \cdot \left(\frac{1}{T_j} - \frac{1}{298}\right)\right) \qquad (3.1)$$

Where A is a constant and T_j is the junction temperature (K). Similarly, the voltage acceleration factor is defined in the *CNET* reliability procedure as [5]:

$$\pi_V = A_3 \cdot \exp\left(A_4 \times V_A \times \left(\frac{T_j}{298} \right) \right) \tag{3.2}$$

Where A_3 and A_4 are constants, V_A is the applied voltage, and T_j is the junction temperature (K). These reliability-prediction models show that the average junction (chip) temperature is a fundamental parameter, and should be accurately estimated for each technology generation. To do this, we must understand the properties of new materials and processes used for implementing VLSIs.

4.   TECHNOLOGY SCALING AND BURN-IN

Traditionally burn-in is used to accelerate the early life of an *IC* to detect the infant mortalities. Figure 3.2 shows the bathtub curve for three different technologies. As we scale to smaller channel lengths, the useful life period of the chip shrinks from more than 7 years (10 years for technologies above 0.25μm) in 0.18μm technology to less than 7 years in 0.10μm technology [6]. This is due to the increasing junction temperature as we scale to deep sub-micron technologies.

The increase in junction temperature arises from higher operating frequencies and consequently higher dynamic power and also increased static power which is due to elevation in leakage power. The useful life period of the *IC* is shrinking due to higher junction temperature operation, higher hot electron injection due to higher I_{Dsat}, and consequently more probable gate oxide wear-out. On the other hand, electro-migration at higher junction temperatures and higher current densities will cause interconnect failures at higher rates. Therefore it is important to carefully optimize the burn-in conditions to avoid over stressing the *ICs* in scaled technologies.

Over stressing the chip in the burn-in environment will further reduce its useful life period and increase post burn-in fallouts. As shown in Figure 3.2, the failure rate in the early stage (0-30 days) of the life of an IC is 500 *DPM* (devices per million) and within the first year it is 200 *FIT*, where 1 *FIT* is 1 failure per 109 hours, or approximately 1 failure per 100,000 years (114,155 years to be precise). The failure rate during the useful operational life of the *IC* is constant but during the aging stage starts to increase due to intrinsic defects (electro-migration, hot electron injection, etc) with a failure rate of less than 0.1%.

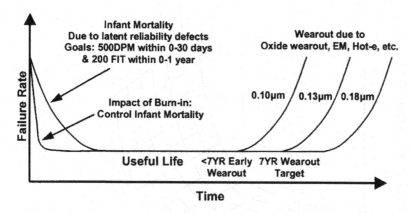

Figure 3-2. Bathtub curve shrinks with technology scaling due to higher junction temperature [53].

4.1 Static Power under Stress Conditions

As we scale the transistor down to the deep sub-micron regime, its off state leakage increases significantly. A linear reduction in transistor threshold voltage with technology scaling results in an exponential increase in its leakage. This leakage is further increased under voltage and temperature stress conditions. The total leakage of a transistor in 0.18µm technology as a function of temperature and voltage stress is shown in Figure 3.3.

The leakage power doubles for every 10°C increase in junction temperature. Since the burn-in test is performed at a reduced frequency (tens of Megahertz), the dynamic power reduces from 75%-80% of total power to a negligible amount when compared to static power.

Figure 3.4 shows that under stress conditions, different leakage components which account for 20% to 25% of the total power under nominal conditions in 0.13µm technology, are increased due to temperature and voltage stress and account for almost all the power under stress conditions. It must be noted that some of these leakage components are mainly sensitive to voltage stress, like gate leakage, and some of them are temperature and voltage sensitive, like sub-threshold leakage.

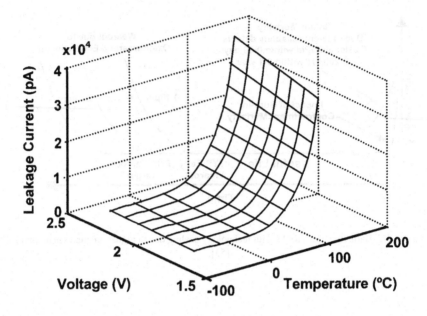

Figure 3-3. SPICE simulation of transistor leakage as a function of voltage and temperature in TSMC 0.18µm technology.

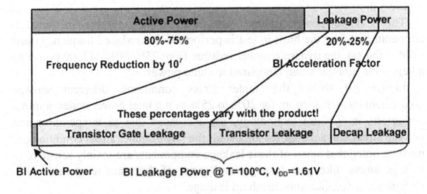

Figure 3-4. Leakage power in burn-in conditions dominates the total power of the chip [6].

5. BURN-IN ELIMINATION

The elimination of burn-in by an alternate screening method has been a long sought after goal. However, despite the expense, mechanical and *EOS/ESD* damage to the burn-in parts, and lengthened time to market, burnt-in parts typically achieve a better quality measure than non-burnt-in parts. The negative features of burn-in stimulated a search for screening methods that might achieve the same lowering of *DPM* levels of shipped parts. In the pre-nano-meter technologies, where transistor channel lengths were above 0.35µm, the I_{DDQ} test was reported by several companies as successful at eliminating or reducing burn-in [7-11].

Intel reported experiments on several thousand *ICs* and found that I_{DDQ}, when combined with a short high voltage stress on the parts, yielded near zero *DPM* outgoing quality levels [7]. Kawasaki Steel reported a similar study using several hundreds of thousands of parts showing that I_{DDQ} screens could eliminate burn-in [8]. LSI Logic and Philips Semiconductors reported similar success with I_{DDQ} screening to eliminate burn-in [9, 10]. McEuen of Ford Microelectronics reported that nominal voltage I_{DDQ} testing enabled reduction of burn-in failures by 51% [11].

However, one caveat of these reports was that I_{DDQ} screening was successful in burn-in elimination only if the manufacturing quality levels were high. I_{DDQ} could not eliminate burn-in on rogue lots. This obstacle was overcome in a study funded jointly by Sandia National Labs and the Sematech organization [12]. The experiment used 3,495 parts in a dynamic burn-in that separated the parts into a control sample, a 7 V stress sample, and an 8 V stress sample. 40,000 I_{DDQ} measurements were taken per die during the control and voltage stress sample tests. I_{DDQ} test limits were set tightly at the +/-3σ levels from the mean plus a tester noise guard-band.

Figure 3.5 summarizes the prediction of functional failure during burn-in from pre-burn-in I_{DDQ} test data. The I_{DDQ} screen predicted that I_{DDQ} testing would detect 50% of the control parts (5 V), 54% of the 7 V stressed parts, and 77% of the 8 V stressed parts. *DPM* of the data showed that the *DPM* level of the control group was 1.75 times larger than the 8 V stressed sample. Cost models also showed economic justification of the I_{DDQ} test in eliminating burn-in. A test methods study was also funded by Sematech with IBM. This is the only study to date that stated that I_{DDQ} testing can not replace burn-in [13]. However, no explanation was given as to why the data contradicted the several reports that it would, and no burn-in data were given.

While these experiments demonstrated that parametric measurements could be used to eliminate burn-in, they were done on long channel transistor *ICs* whose background noise levels obscured sensitive I_{DDQ} or other parametric measurements. An important question now is how does I_{DDQ} or other parametric measurements perform for nano-meter *CMOS ICs*. There are two public reports of success. The first was at a burn-in panel at the International Reliability Physics Symposium (*IRPS*) in 2001 [14]. Panelists from five major companies said that if the manufacturing quality of the lots could be measured as high, then parametric screens could achieve *BI* elimination. They stressed that this approach did not work if the quality levels were not high.

The second report on nano-technology parts came from a team from LSI Logic and Portland State University [15-17]. They reported parametric screening of outlier parts using post-test statistical processing methods on the whole wafer data. The technique measures statistics of neighboring or other die locations on the wafer to determine I_{DDQ} and V_{DDMin} (lowest functional voltage V_{DD}) test limits. These studied reported the application of post-test statistics to burn-in elimination, but did not specifically report the burn-in elimination data. The severe problems that nano-meter *ICs* present to burn-in, make these parametric screening techniques of great interest.

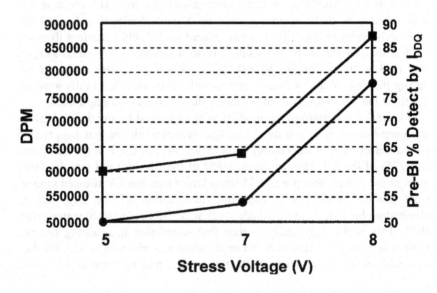

Figure 3-5. I_{DDQ} detection of burn-in functional failures and the defect level of *ICs* that failed only I_{DDQ} tests [12].

6. ESTIMATION OF JUNCTION TEMPERATURE INCREASE WITH TECHNOLOGY SCALING UNDER BURN-IN CONDITIONS

The burn-in screening procedure weeds out latent defects from a product, and thereby improves the outgoing quality and reliability of the product. During burn-in, *ICs* are subjected to elevated temperature and voltage in excess of normal operating conditions for a specific period of time. This accelerates the product lifetime through the early part of its life cycle allowing removal of the products that would have failed during that time.

There are die level burn-in (*DLBI*) and wafer level burn-in (*WLBI*) techniques. *DLBI* can handle, contact, and do burn-in stress on several packaged die together, while *WLBI* has the ability to contact every die location and perform the burn-in test simultaneously on an entire wafer. For the *DLBI*, one must also consider the thermal impedance network of the package [18]. Once this network is known, then Eq. (3.3) can be suitably modified to reflect the total thermal resistance (R_{ja}) of the die and many types of package. In this work, the focus was on the intrinsic behavior (junction temperature estimation) of the silicon die under burn-in conditions for the sake of simplicity. In other words, the thermal impedance network of the package is not considered.

The average inverter power for different operating conditions and technologies (Table 3.1) was estimated by simulating the inverters at different temperatures and V_{DD}. For burn-in, the stress temperature was varied from 25°C to 125°C. Similarly, the stress voltage was varied from nominal V_{DD} for the given technology to $V_{DD}+30\%$, and in this simulation (BSIM model level 49) the inverter input was grounded. The simulated I_{av} and the calculated values of P and ΔT are given in Table 3.1, where I_{av} and P are the average current and power dissipation of an inverter, and ΔT is (T_j - T_a) per $1mm^2$ of chip area calculated using Eq. (3.3).

$$\Delta T = P_{transistor} \times R_{ja-transistor} \times \frac{D_{density}}{2} \left[\frac{°C}{mm^2} \right] \qquad (3.3)$$

Where $P_{transistor}$ is the power dissipation of the off-mode transistor in the inverter, $R_{ja-transistor}$ is the thermal resistance of the on-transistor in the inverter, and $D_{density}$ is the transistor density in the *CMOS* chip. For a given technology, the thermal resistance was extracted from Figure 2.15 and the transistor density was calculated from Figure 2.17. It was assumed that the circuit under study is a fully static *CMOS* design. Therefore, half of the total transistors are in the off-mode during burn-in, and this was taken into account by dividing $D_{density}$ by 2 in Eq. (3.3).

Table 3-1. DC simulation (I_{av}) and calculation results (P, ΔT) of CMOS inverters for different technologies.

CMOS Technology		25°C			85°C			125°C		
		I_{av}, pA	P, pW	ΔT,°C /mm²	I_{av}, nA	P, nW	ΔT,°C /mm²	I_{av}, nA	P, nW	ΔT,°C /mm²
0.35 µm	3.30V	7.70	25.0	71E-5	0.070	0.23	66E-4	2.05	6.77	0.20
	3.80V	9.20	35.0	99E-5	0.084	0.32	91E-4	2.15	8.17	0.23
	4.30V	11.1	47.7	14E-4	0.110	0.47	14E-3	2.27	9.76	0.28
0.25 µm	2.50V	19.3	48.3	23E-4	0.418	1.04	0.050	3.96	9.90	0.29
	2.90V	22.0	63.8	31E-4	0.470	1.36	0.065	4.41	12.80	0.35
	3.25V	25.0	81.3	39E-4	0.531	1.75	0.080	4.81	15.87	0.45
0.18 µm	1.80V	90.5	163	0.020	1.330	2.39	0.24	8.96	16.13	0.97
	2.10V	101	210	0.022	1.480	3.08	0.31	9.85	20.48	1.23
	2.35V	112	264	0.027	1.620	3.81	0.39	10.9	25.60	1.51
0.13 µm	1.20V	766	920	0.200	8.45	10	2.32	28	34	7.79
	1.40V	1200	1680	0.380	12.3	17	3.94	34	47	10.97
	1.56V	1760	2900	0.670	17.45	27.6	6.40	55	85	19.81

Table 3-2. N-MOSFET parameters used for simulations.

	UHP	LP
Substrate doping, cm⁻³ (p - type)	5×10^{15}	5×10^{15}
Source/Drain doping, cm⁻³ (n - type)	3×10^{20}	3×10^{20}
V_{TH} adjusted doping, cm⁻³ (p - type)	1.8×10^{18}	3×10^{18}
Punch - Trough doping, cm⁻³ (p - type)	5×10^{19}	8×10^{19}
Effective gate oxide thickness, Å	18	18
L_{eff}/W, nm/µm	63/2	63/2
Nominal $V_{DS} = V_{DD}$, V	1.0	1.0

In this section, *UHP* and *LP* devices were considered as worst and best cases with respect to power consumption during burn-in. The transistor parameters obtained from simulations under normal operating conditions are presented in Table 3.2. The dominant components of the leakage current in a sub-100 nm *MOSFET* are sub-threshold, band-to-band tunneling, and gate oxide tunneling currents [19].

Since there was no access to industrial *HSPICE* device models for the 90 nm *CMOS* technology, the *HSPICE* simulations in Cadence for this technology generation could not be used. To predict the possible increase of average junction temperature in *CMOS* chips under burn-in conditions, an *NMOSFET* at stressed operating conditions was simulated using the 2-D device simulator "Microtec" [20]. The *MOSFET* parameters used for device simulations are given in Table 3.3. The simulation results correspond to *DC* characteristics of 90 nm transistors [21, 22, 23], such as $V_{TH} = 0.2 - 0.28V$, $I_{ON} = 600\text{-}750\mu A/\mu m$ and $I_{OFF} =20\text{-}100nA/\mu m$. These devices were developed for ultra high performance applications (*UHP*). Low power (*LP*) medium speed [21, 23] devices assume $V_{TH} = 0.3\text{-}0.35V$, $I_{ON} = 480\text{-}520 \ \mu A/\mu m$ and $I_{OFF} =0.18\text{-}0.5nA/\mu m$. High performance (*HP*) applications assume a leakage current of approximately *10nA/μm* [24].

Table 3-3. DC parameters for an *N-MOSFET* emulated in 90-nm *CMOS* technology (V_{DD} = 1V, T = 25°C).

	L_{eff}, nm	V_{TH}, V	I_{ON}, uA/um	I_{OFF}, nA/um
UHP	63	0.25	600	30
LP	63	0.35	440	0.6

The simulation results of an averaged sized *MOSFET* (*W/L* $=2.0\mu m/0.1\mu m$) under stressed operating conditions are given in Table 3.4. In this table, *P* is the power dissipation of an off-mode inverter transistor that was obtained from device simulations. *ΔT* (thermal density) is the ($T_j\text{-}T_a$) per *1mm²* of *CMOS* chip that was calculated using Eq. (3.5). The transistor density in a *CMOS* chip was assumed to be *0.27 millions/mm²* (Figure 2.17). When *ΔT* in Table 3.1 and Table 3.4 was calculated, it was assumed that each off-mode transistor in a *1mm²* chip area was an independent heat source. The total junction temperature increase of this area over ambient temperature was defined as the product of heat source density and the junction temperature increase of a single transistor. In practice, the thermal coupling effect of transistors on a chip must be considered, and this depends on layout.

In the first order approximation, the thermal coupling effect of transistors was neglected in this analysis. Table 3.1 and Table 3.4 show that the average leakage current and dissipated power is increased by at least two orders of magnitude by technology scaling if the ambient temperature is 85°C or less. At 125°C, the increase in current and power dissipation with technology scaling is relatively less. However, the increase in ΔT is more dramatic owing to increased transistor density, leakage current, and the thermal resistance.

Table 3-4. Predicted power dissipation and junction temperature increase in a *CMOS* inverter (90nm *CMOS* technology).

	LP	LP	LP	UHP	UHP	UHP
V_{DD} (V)	1.3	1.15	1.0	1.3	1.15	1.0
P, pW, -100°C	0.34	0.208	0.124	0.120	0.084	0.057
ΔT,°C/mm² -100°C	1.5×10^{-4}	9.1×10^{-5}	5.4×10^{-5}	0.052	0.036	0.021
P, nW, 0°C	0.75	0.51	0.32	44.2	29	18.4
ΔT,°C/mm² 0°C	0.33	0.23	0.14	19.3	12.66	8.03
P, nW, 25°C	2.7	1.84	1.2	130	82.8	60
ΔT,°C/mm² 25°C	1.18	0.81	0.53	56.8	36.15	26.2
P, nW, 85°C	21.23	14.63	9.8	770	506	328
ΔT,°C/mm² 85°C	9.27	6.39	4.28	336.2	221	143.2
P, nW, 125°C	152.9	107.6	74.4	3084	2047	1344
ΔT,°C/mm² 125°C	66.75	46.99	32.49	1346.6	893.8	586.8

The normalized temperature increase of a *CMOS* chip with scaling at burn-in conditions is shown in Figure 3.6. The plot with diamond symbols depicts the normalized T_j increase if $T = 125°C$. For 90nm technology the increase in T_j is different for the high performance or low power process. If all the transistors are implemented with low V_{TH} UHP devices (unrealistic) then the normalized T_j is increased by approximately 5000x compared to 0.35μm *CMOS*. On the other hand, if all the transistors are implemented with *LP* devices, then the T_j is increased by approximately 230x. It should be noted that most of the transistors on a chip will be implemented with *LP* devices. However, if the T_a is reduced by 10°C for each technology generation, the normalized T_j is also reduced as shown by the plot with square symbols. Similarly, leakage reduction techniques can also be employed to further reduce the increased normalized temperature with scaling [25, 26]. If such techniques are employed as well as the T_a being reduced by 10°C for each technology generation, the normalized T_j increase for 90 nm *CMOS* with respect to 0.35μm *CMOS* becomes relatively small (7-8x).

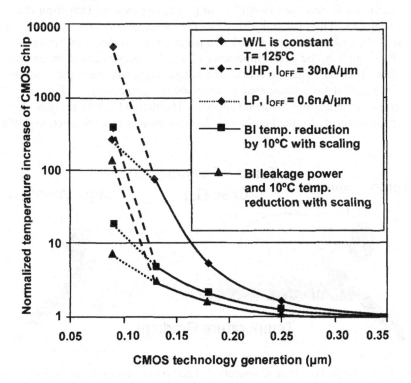

Figure 3-6. Normalized chip junction temperature at V_{DD} + 30% burn-in condition.

In spite of the reduction in T_a and the use of leakage reduction techniques, the increase in T_j is still clearly unacceptable. Obviously, burn-in conditions should be carefully optimized for 130 nm and 90 nm *CMOS* technologies to reduce the risk of chip over stressing during burn-in.

7.   PACKAGING CONSIDERATION IN BURN-IN

High performance VLSI circuits, such as microprocessors, significantly challenge power delivery and heat removal due to smaller dimensions and increasing power dissipation. Technical challenges in the thermal management of microprocessors arise from two causes [27]:
- Increased dynamic and leakage power dissipation associated with technology scaling.
- Heat removal from localized hot spots.

The former is especially important for burn-in since the leakage power is exponentially increased under stress conditions. Typically, thermal management features are integrated in packages to spread heat from die to the heat sink. The heat sink dissipates the heat into local environments. A typical thermal resistance network of a packaged die is shown in Figure 3.7.

By definition, the case temperature (T_c) is the temperature at the external surface of the package. All semiconductor packages have multiple elements. In the simplest form these elements include the semiconductor die, thermal interface material, and the heat sink base. The thermal conductivity of these package elements for the Pentium III Xeon microprocessor is given in Table 3.6.

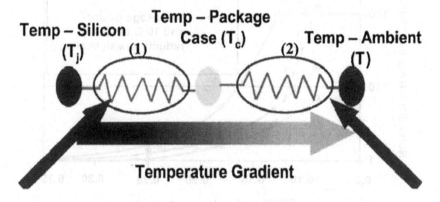

Figure 3-7. Thermal resistance network of a packaged die: (1) junction to case (package), (2) case to ambient (heat sink) [27].

In a common case, the junction temperature increase over ambient temperature has three components [28]:

$$\Delta T = P \times \left[R_{th(die\text{-}pack)} + R_{th(pack-\sin k)} + R_{th(\sin k-amb)} \right] \qquad (3.4)$$

Table 3-5. Thermal conductivity of package components [32].

Packaged Components	Conductivity, W/mK
Silicon die	123
Thermal interface material	3.8
Heat sink base	180

Where $R_{th(Die\text{-}pack)}$, $R_{th(Pack\text{-}sink)}$, $R_{th(Sink\text{-}amb)}$ are the die to package, package to heat sink, and heat sink to ambient thermal resistances, respectively, and P is the total power dissipation of the chip. The first component in Eq. (3.4) is discussed in previous sections. The third component is determined by the cooling techniques and will be considered in the next section. Here, the second component in Eq. (3.4) is considered and can be rewritten as follows:

$$\Delta T_{(Package)} = 0.5 \times P_{MOSFET} \times D \times R_{th(Pack\text{-}sink)} \qquad (3.5)$$

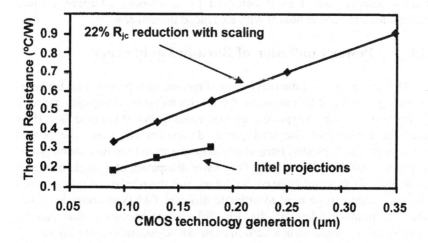

Figure 3-8. Reduction of package thermal resistance with technology scaling.

In Eq. (3.5), P_{MOSFET} is the transistor power dissipation and D is the transistor density. The package to heat sink thermal resistance, $R_{th}(pack\text{-}sink)$, is crucial to removing heat during burn-in. Values of 0.9-1.2°C/W were reported for $R_{th}(pack\text{-}sink)$ in 350 nm technology [29, 30]. It is predicted that a reduction of approximately 22% in $R_{th}(pack\text{-}sink)$ per technology generation is required to just compensate for the increased power density with technology scaling [31]. Figure 3.8 shows these projections for the 350 nm technology to 90 nm technology.

8.   COOLING TECHNIQUES FOR BURN-IN

Low power devices can be burnt-in without attention to thermal considerations. However, as power dissipation increases with technology scaling for high performance chips, burn-in requires advanced cooling concepts and additional hardware to facilitate direct contact between the heat sink and the die. Advanced burn-in ovens should provide uniform temperature distribution in the chamber and precise temperature control for each individual device. The power dissipation within one lot of devices can vary by 40% due to manufacturing variations and different test vectors applied during burn-in. This variation in power, and approximately 30% variation in oven airflow, can create a significant variation in package temperature [33]. If the device becomes too hot, it may be damaged while other devices may not be adequately burnt-in. To uniformly stress all devices, each package device temperature must be kept close to the specified burn-in temperature. This is achieved by developing advanced cooling techniques and burn-in boards with embedded thermal sensors.

8.1   Power Limitation of Burn-in Equipments

The total number of die that can be simultaneously powered-up for burn-in testing will likely be limited by the maximum power dissipation capacity of the burn-in oven. A typical oven may contain several hundred dies. If all dies are active, then the total power dissipation can reach the several kilowatt range. Typically, burn-in ovens have a maximum dissipation power between 2500-6500 Watts [34]. The power dissipation of a single transistor in an inverter in static stressed conditions and the number of transistors of the logic chip can be used to estimate different *CMOS* technologies. Then the maximum number of die for different technologies that can be simultaneously powered in a burn-in oven can be estimated using Eq. (3.6):

$$N_{dies} = \frac{P_{oven}}{P_{transistor} \times \frac{N_{transistors}}{2}} \tag{3.6}$$

Where P_{oven} is the maximum power dissipation of the burn-in oven at stressed conditions, $P_{transistor}$ is the power dissipation of a single transistor at static stressed conditions for the given technology, and $N_{transistors}$ is the total number of transistors in the logic chip for the given technology. Eq. (3.6) assumes that 50% of the total numbers of transistors are off at any point during burn-in assuming fully static *CMOS* design. Results are shown in Figure 3.9. Burn-in ovens, such as the *PBC1-80* of Dispatch Industries [34] and Max-4 of Aehr Test Systems [1] have maximum power dissipation of about 2500 and 15,000 watts, respectively, at 125°C. The room ambient temperature is assumed to be 25°C.

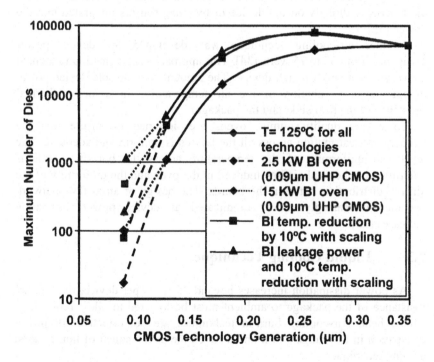

Figure 3-9. Maximum number of dies for one burn-in load with scaling.

8.2   Air Cooling Technique

For *CMOS IC* technologies of 0.35μm and above, generally *IC* junction heating during burn-in has not been a major issue and the oven temperature could be easily set to avoid temperature-related over stress. However, for 0.25μm technology and below, device self-heating has been described to become a more significant issue and air-cooling techniques began to be implemented to remove heat from each device and the oven.

Air-cooled burn-in ovens are reasonably effective in heat removal from devices dissipating up to 30-40 watts [33]. Often, an air-cooled heat sink and embedded thermal sensors are used to control the individual temperature of each device. The air temperature and air velocity are dependent on the device power, the overall thermal resistance of the heat sink assembly and burn-in socket, and the required package temperature. The air temperature and velocity must be controlled so that the embedded heat sink can limit the device temperature increase over the range of heat dissipation. The device temperature can be controlled in the range of 50°C-150°C with an accuracy of 3°C [33]. Device temperature is usually measured by attaching a small thermocouple directly on the device or by using sensors integrated into the device [35].

Another air-cooling technique was developed for device power dissipation from 35 to 75 watts [34]. This approach uses a small fan mounted above the heat sink of each device. The amount of allowable device power dissipation is a function of the air temperature, air velocity, thermal resistance of the heat sink, and the package.

To ensure quality output, ovens are designed to ensure that the temperature distribution across all the boards is uniform and adequate. The level and uniformity of the temperature across the burn-in boards is controlled by the total airflow induced in the oven and the uniformity of the airflow distribution between the boards. The design of an airflow network becomes increasingly more complicated as device power dissipation increases [36].

8.3   Liquid Cooling Technique

As power dissipation increases beyond 75 watts per device, the thermal resistance of the package to ambient must be lowered to allow removal of excess heat. Air-cooling burn-in techniques are not effective for power dissipation in this range and it has fostered the development of liquid-based cooling techniques.

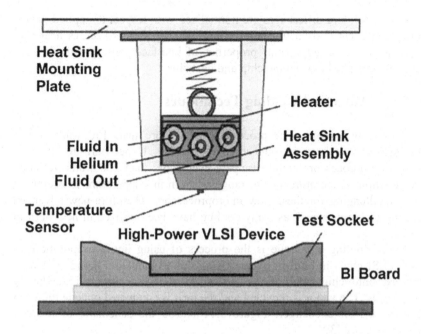

Figure 3-10. Water-cooled heat sink, adopted from [34].

Figure 3.10 illustrates one such technique [34]. A temperature sensor embedded in the heat sink measures the device temperature. Helium is injected into the heat sink to provide a lower thermal interface between the device and the heat sink. This technique lowers the heat sink to ambient thermal resistance by approximately 40%.

Each heat sink has a temperature-controlled heater. The burn-in ovens with liquid-cooled heat sinks can burn-in devices that dissipate over 150 W of power [37,38]. In such ovens, the ambient temperature for each device can be optimized for optimal burn-in conditions. This is important since self-heating dissipation can vary significantly due to inherent process spreads in scaled technologies.

The thermal control during test and burn-in of devices with high leakage power dissipation (above 75 Watts) plays a key role in increasing the post burn-in yield. Special thermal test chips and modules were developed to measure temperature gradients in packages and heat sinks in burn-in equipment [37, 38]. For example, IBM used a *TV994* thermal test chip for burn-in equipment qualification. This $14.7mm^2$ chip has nine small resistive temperature detectors (*RTD*) and four large heater resistors, one covering each quadrant of the chip [37]. The thermal interface tests evaluate temperature gradients within the device and between the device and heat sink. Temperature differences are normalized with respect to applied device power. The test is used to optimize and evaluate factors such as heat sink material, flatness and various properties of interface pads, and liquids and gases that can be between the chip and heat sink.

8.4   Advanced Cooling Techniques

In last two years microprocessors with more than 150 watts power dissipation have been introduced to the market. Under burn-in condition these microprocessors could dissipate even more than 200 watts. Keeping the junction temperature in the range of burn-in temperature is becoming more challenging for these new microprocessors. Therefore new advanced cooling techniques such as spray cooling have been utilized in new burn-in ovens.

Spray cooling technique is the process of using liquid evaporation, to cool VLSI chips in burn-in ovens. Using this technique, a liquid coolant is sprayed onto chips and due to hot surface of the chip it immediately evaporates. The vapor is recycled and the heat is conducted outside the oven through a heat exchanger. Figure 3.11 shows the schematic of a spray cooling system [39].

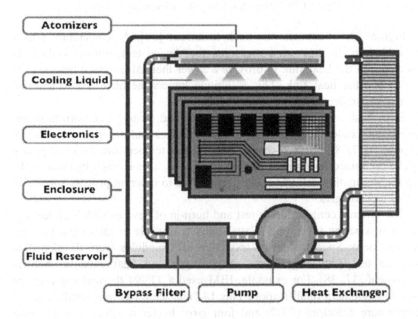

Figure 3-11. The schematic of the spray cooling system used to cool VLSI chips with high power dissipation [39].

A non-conductive and non-corrosive coolant is atomized and sprayed directly onto chips to provide cooling. The coolant vaporizes and heat is rejected to the enclosure and/or through a heat exchanger, condensing the vapor back into a liquid state. The process continuously cycles within a closed loop sealed enclosure that prevents corrosive environmental contamination from harming sensitive electronics. The junction to ambient thermal resistance of the chip in burn-in ovens using spray cooling can be as low as 0.25°C/W.

9.   BURN-IN LIMITATIONS AND OPTIMIZATION

Yield and reliability are two important factors in semiconductor manufacturing. Typically three parameters significantly affect the yield and reliability of *ICs* [40]:
- Design-related parameters (chip area and gate oxide thickness).
- Process-related parameters (defect distribution and density).
- Operation-related parameters (voltage and temperature).

It has been experimentally verified that defects that cause burn-in failures (early-life reliability failures) are fundamentally the same in nature as defects that cause wafer probe failures (yield failures) [41, 42]. Researchers have also identified two key reliability indicators in order to optimize yield during burn-in:
- Local region yield.
- The number of defects that have been repaired (for chips containing redundancy).

Experimentally, it has been shown that die with many faulty neighbors can pose a significantly greater early-reliability risk than chips with few faulty neighbors [43]. An *IC* with a redundancy-related repair is more likely to have a latent defect mechanism resulting in early life failure [41].

The key to optimizing burn-in lies in identifying those die that most likely to fail during burn-in before the burn-in is actually performed. Once identified, die of higher reliability risk may be subjected to more rigorous testing (longer burn-in duration), while those dies deemed more reliable may have a reduced stress, or no stress at all. Barnett et al. proposed the post burn-in yield model, which include the burn-in time as a parameter [43]. It was assumed that the average number of latent defects (λ_L) per chip is time-dependent as follows:

$$\lambda_L(t) = \alpha.\gamma.(1 - Y_K^{1/\alpha}).(\frac{t}{\tau})^\beta \qquad (3.7)$$

Where α is the defect clustering parameter, $\gamma = 0.01\text{-}0.02$ is the fitting parameter, Y_K is the wafer test yield (yield before burn-in), t is the burn-in time in hours, and β is the shape parameter of the Weibull distribution of the reliability function. The post burn-in reliability yield (i.e. the number of dies surviving burn-in) is modeled as follows:

$$R(t) = \left[1 + \frac{\lambda_{I_r}(t)}{\alpha}\right]^{-\alpha} \tag{3.8}$$

Kim et al. [44] developed another model for post burn-in reliability (R) and yield loss (Y_{loss}), which will be discussed later in this section.

Burn-in removes the infant mortality device population hence improving the outgoing device reliability. However, burn-in may affect the post burn-in yield of *ICs* since latent defects may become enhanced during burn-in, with a resultant increase in post burn-in yield loss. The amount of yield loss depends on burn-in conditions (voltage, temperature, time). Since the stress voltage and the stress temperature provide the acceleration during burn-in, the burn-in time is the parameter that is manipulated to control the post burn-in yield loss using above mentioned models. In practice, many *IC* manufactures reduce the burn-in time to 10 hours or even skip burn-in, when the yield before burn-in is high (~98%) and burn-in escapes is low (~100 *PPM*) [47]. The amount of burn-in escape is estimated by the early failure rate test, which is performed on 10000 final products from at least three lots with duration approximately 12-48 hours under burn-in conditions.

Several reliability failure mechanisms are accelerated by temperature, so burn-in testing is done at elevated temperature. These mechanisms include metal stress voiding and electro-migration, metal slivers bridging shorts, contamination, and gate-oxide wear out and breakdown [12]. However, there are physical and burn-in equipment related limitations for temperature and voltage stress. Die failure rate (failures per million) increases exponentially with temperature for most failure mechanisms [45]. As a result, the yield loss may increase if the burn-in conditions are over stressed. Hence, we should optimize the junction temperature of die for normal and burn-in conditions.

9.1　　Physical and Practical Limits of Junction temperature

The maximum operating temperatures for semiconductor devices can be estimated from the semiconductor intrinsic carrier density that depends on the band-gap of the material. When the intrinsic carrier density reaches the doping level of the active region of devices, electrical parameters are expected to change drastically. The highest reported operating junction temperature is about 200°C in standard silicon technology [46]. At this temperature, the circuit performance is reduced substantially. The temperature will affect thermal conductivity, built-in potential, threshold voltage, and *pn* junction reverse current. Several practical considerations limit the junction temperature to a much lower value. A value of 150°C for junction temperature is often used for ICs as the limit [47].

The peak junction temperature in a PowerPC microprocessor implemented in a 0.35µm *CMOS* technology with a 0.3µm effective transistor channel lengths is about 90°C-100°C at an operating speed of 200-250 MHz [48, 49]. If this is used as the reference temperature and assuming that Figure 3.4 estimates the junction temperature increase with reasonable accuracy and package thermal resistance remains the same, then one can expect a 2.4x increase in junction temperature for the same microprocessor implemented in a 0.18µm *CMOS* technology. Hence, the die junction temperature should be approximately 156°C-180°C.

These values are obtained assuming cooling, packaging and circuit techniques remain the same when moving from 0.35µm technology to 0.18µm m technology. However, improved cooling and packaging considerations will reduce the temperature to a much lower value. Similarly, circuit techniques such as transistor stacking, dual-threshold transistors, reverse body bias, etc. can reduce substantially leakage current and the junction temperature.

9.2 Optimization of Burn-In Stress Conditions with Technology Scaling for Constant Reliability

The optimal burn-in conditions for maintaining the projected failure rate require that the defect distribution models and their growth models be studied. The post burn-in reliability (*R*) and yield loss (Y_{loss}) have been studied [44, 50]. T. Kim, et al., [44] proposed the following models for post burn-in reliability and yield loss shown in Eq. (3.9) and Eq. (3.10):

$$Y_{loss} = Y.(1 - Y^{\frac{v}{1-v}}) \qquad\qquad (3.9)$$

$$R = Y^{\frac{1}{(1-u)^2 - 1}} \qquad\qquad (3.10)$$

Where Y is the yield before burn-in, and u, v are constants that depend on the stress temperature and voltage. Using the I/E gate oxide breakdown model and the post burn-in yield loss model, Vassighi, et al., demonstrated that the post burn-in yield loss increases exponentially with increasing stress temperature for a given stress voltage [50]. This result was obtained for a $0.18\mu m$ *CMOS* technology ($T_{ox} \approx 41A^{\circ}$).

Hence, over stressing a die during burn-in may significantly reduce the post burn-in reliability and increase the yield loss, especially when the junction temperature at burn-in and normal operating conditions are increased with technology scaling. Thus, to a first order, we want a constant reliability during burn-in with technology scaling.

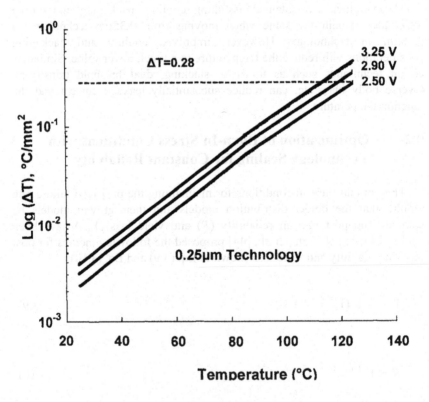

Figure 3-12. ΔT as a function of ambient temperature and V_{DD} for $0.25\mu m$ technology.

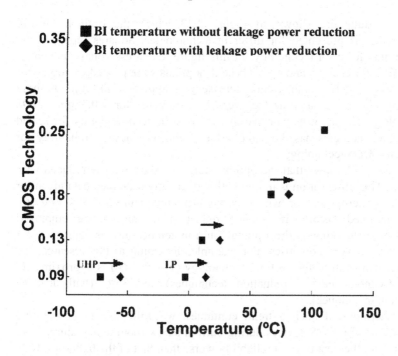

Figure 3-13. Optimal burn-in temperature for constant reliability.

The burn-in temperature and voltage should be optimized for different *CMOS* technologies to maintain the average junction temperature of the die at the same fixed level. If electrical defect densities are equal, then we assume that the post burn-in reliability for an advanced *CMOS* technology should not be worse than the post burn-in reliability for the 0.35μm *CMOS* technology. This means that the junction temperature increase over ambient temperature during burn-in for subsequent technologies should not be higher than the burn-in junction temperature increase for 0.35μm *CMOS* technology. Table 3.1 shows that for 0.35μm *CMOS* technology, the junction temperature increase (ΔT) over ambient stressed temperature per mm^2 of chip is 0.28°C at $V_{DD} = 4.3V$, $T = 125°C$.

The horizontal line on Figure 3.12 illustrates this limit. Now for 0.25μm technology, if the $\Delta T/mm^2$ versus stress temperature is plotted for three different stress voltages, it results in three different curves. Subsequently, the optimal burn-in temperature where the horizontal line ($\Delta T= 0.28°C/mm^2$) intersects with graphs can be found. Similarly, the optimal burn-in temperature for other technologies can be found using data from Table 3.1 and Table 3.4.

The results are shown in Figure 3.13 where the optimal burn-in temperature is presented for different technologies. Squares represent the data points for each technology. In this figure, the stress voltage is kept at $V_{DD}+30\%$ for each technology. These data points were plotted ensuring that the average junction temperature increase over ambient (ΔT) for die in these technologies is the same as the average ΔT increase for a 0.35µm *CMOS* technology. Hence, it is expected that the post burn-in reliability for scaled *CMOS* technologies has the same value as the post burn-in reliability for 0.35µm *CMOS* technology.

Figure 3.13 shows that the optimal burn-in temperature is reduced with scaling. This observation is in line with the recently presented data given for a 0.18µm microprocessor, where the burn-in temperature is 85°C- 90°C [6]. As mentioned before, if leakage reduction techniques are employed (diamond data points), the optimal burn-in temperature is increased for 0.18µm or lower geometries. For example, according to this research, the optimal temperature for 130 nm technology ($V_{DD} \gg 1.4V$) is approximately 10°C (without leakage reduction techniques) and 35°C (with leakage reduction techniques).

Furthermore, if such a trend continues, we will have to cool future generations of *CMOS* devices during burn-in below room temperature, if we do not want the post burn-in reliability worse than that of the 0.35µm *CMOS* technology. For example, the estimated burn-in temperature for a 90nm *CMOS* technology may be approximately 0°C to 15°C. Note, that many future chips will use a mixture of technologies: *UHP* logic is for critical delay paths and *LP* logic is for the low activity *SRAM* cells [29].

10.   SUMMARY

The impact of technology scaling on the burn-in environment was investigated. The following conclusions are obtained: Firstly, there is a steady increase in the junction temperature with scaling. The normalized junction temperature increase under burn-in conditions becomes exponential with technology scaling if no leakage reduction techniques are used. On the other hand, if leakage reduction techniques are used, then an approximately linear increase in junction temperature can be obtained. As a consequence, the burn-in temperature must be linearly reduced with scaling in order to maintain same junction temperature across the technology nodes. Second, the number of dies that can be simultaneously burnt-in is reduced with technology scaling, because of the maximum power dissipation limit of presently available burn-in ovens. Finally, the optimal stressed temperature in a burn-in environment is significantly reduced with technology scaling.

It was also argued that deep sub-micron devices will require advanced packaging and liquid cooling techniques to lower the junction to ambient thermal resistance. In scaled technologies, burn-in optimization for yield and reliability is of crucial significance owing to larger number of design and technology variables. In some situations, individual chip level burn-in optimization will be necessary in order to provide optimum burn-in environment for each chip.

References

1. Aehr Test Systems. www.aehr.com.
2. Micro Control Co. http://www.microcontrol.com/.
3. Despatch Industries. http://www.despatch.com/pdfs/PBC.pdf.
4. P. Lall. "Tutorial: temperature as an input to microelectronics-reliability models". IEEE Trans. on Reliability, 45(1):3-9, 1996.
5. J.B. Bowles. "A survey of reliability-prediction procedures for microelectronics devices". IEEE Trans. on Reliability, Vol. 41, No. 1, pages 2-12, 1992.
6. T. M. Mak. "Is CMOS more reliable with scaling?" www.intel.com/technology/itj/q11999/articles/art-6who.htm.
7. T. Henry and T. Soo. "Burn-In Elimination of a High Volume Microprocessor Using IDDQ". IEEE International Test Conference (ITC), pages 242-249, 1996.
8. R. Kawahara, O. Nakayama, and T. Kurasawa. "The Effectiveness of IDDQ and High Voltage Stress for Burn-in Elimination". IEEE IDDQ Workshop, pages 9-14, 1996.
9. T. Barrette, V. Bhide, K. De, M. Stover, and E. Sugasawara. "Evaluation of Early Failure Screening Methods". IEEE IDDQ Workshop, pages 14-17, 1996.
10. K. Wallquist. "On the Effectiveness of ISSQ Testing in Reducing Early Failure Rate". International Test Conference (ITC), pages 910-915, 1995.
11. S. McEuen. "Reliability Benefits of IDDQ". Journal of Electronic Testing: Theory and Applications (JETTA), Vol. 3, No. 4, pages 327-335, 1992.
12. A.W. Righter, C.F. Hawkins, J.M. Soden, and P. Maxwell. "CMOS IC reliability indicators and burn-in economics". Proceeding of International Test Conference, pages 194-203, 1998.

13. P. Nigh, D. Vallett, P. Patel, J. Wright, F. Motika, D. Forlenza, R. Kurtulik, and W. Chong. "Failure analysis of timing and IDDQ-only failures from the SEMATECH test methods experiment". International Test Conference. (ITC), pages 43-52, 1998.

14. "Panel on Burn-in Elimination". International Reliability Physics Symposium (IRPS), 2001.

15. R. Daasch, K. Cota, J. McNames, and R. Madge. "Neighbor selection for variance reduction in IDDQ and other parametric data". International Test Conference. (ITC), pages 92-100, 2001.

16. R. Madge. "Screening Min VDD Outliers Using Feed-Forward Voltage Testing". International Test Conference. (ITC), pages 673-682, 2002.

17. C. Schuermyer, B. Benware, K. Cota, R. Madge, R. Daasch, and L. Ning. "Screening VDSM Outliers Using nominal and Subthreshold Supply Voltage IDDQ". International Test Conference. (ITC), pages 565-573, 2003.

18. G. Kromann. "Thermal management of a C4/CBGA interconnect technology for a high-performance RISC microprocessor: The Motorola PowerPC 620TM microprocessor". Proceedings of IEEE Electronic and Technology Conference, pages 652-659, 1996.

19. D.J. Frank. "Power-constrained CMOS scaling limits". IBM Journal of Research and Development, Vol. 46, No. 2/3, pages 235-244, 2002.

20. Siborg Corp. web-site: http://www.siborg.com/.

21. K. Fukasaku, A. Ono, T. Hirai, Y. Yasuda, N. Okada, S. Koyama, T. Tamura, Y. Yamada, T. Nakata, M. Yamana, N. Ikezawa, T. Matsuda, K. Arita, H. Nambu, A. Nishizawa, K. Nakabeppu, and N. Nakamura, "UX6-100 nm generation CMOS integration technology with Cu/Low-k interconnect", Proceedings of Symposium on VLSI Technology, pages 64-65, 2002.

22. S.F. Huang, C.Y. Lin, Y.S. Huang, T. Schafbauer, M. Eller, Y.C. Cheng, S.M. Cheng, S. Sportouch, W. Jin, N. Rovedo, A. Grassmann, Y. Huang, J. Brighten, C.H. Liu, B.V. Ehrenwall, N. Chen, J. Chen, O.S. Park, and M. Common, "High-performance 50 nm CMOS devices for microprocessors and embedded processor core applications", IEDM, pages 237-240, 2001.

23. A. Ono, K. Fukasaku, T. Hirai, S. Koyama, M. Makabe, T. Matsuda, M. Takimoto, Y. Kunimune, N. Ikezawa, Y. Yamada, F. Koba, K. Imai, and N. Nakamura, "A 100 nm node CMOS technology for practical SOC application requirement". IEDM, pages 511-514, 2001.

24. S. Thompson, P. Packan, and M. Bohr. "MOS scaling: transistor challenges for the 21st century". Intel Technology Journal, Vol. Q3, pages 1-19, 1998. http://developer.intel.com/technology/itj/archive.htm.

25. S. Borkar. "Leakage reduction in digital CMOS circuits". Proc. of IEEE Solid-State Circuits Conference, pages 577-580, 2002.
26. A.B. Kahng. "ITRS-2001 design ITWG". ITRS Release Conference.
27. R. Mahajan, R. Nair, V. Wakharkan, J. Swan, J. Tang, and G. Vandentop. "Emerging directions for packaging technologies". Intel Technology Journal, Vol. 6, No. 2, pages 62-75, 2002. http://developer.intel.com/technology/itj/2002/volume06issue02/.
28. J.W. Worman. "Sub-millisecond thermal impedance and steady state thermal resistance explored". Proceedings of IEEE SEMI-THERM Symposium, pages 173-181, 1999.
29. "Pentium Processor with MMX Technology". http://cs.mipt.ru/docs/comp/eng/hardware/processors/intel/i586/p55/main. pdf.
30. "IBM 6X86MX Microprocessor". http://www3.ibm.com/chips/techlib/techlib.nsf/techdocs/AF16346AD95E 76D987256A310064E3B4.
31. K. Banerjee and R. Mahajan. Intel Development Forum, 2002. ftp://download.intel.com/research/silicon/Thermals-press-IDF-0902.pdf.
32. T.J. Goh, K.N. Seetharamu, G.A. Quadir, and Z.A. Zainal. "Thermal methodology for evaluating the performance of microelectronic devices with non-uniform power dissipation". Proceedings of IEEE Electronics Packaging Technology Conference, pages 312-317, 2002.
33. H.E. Hamilton. "Thermal aspects of burn-in of high power semiconductor devices". IEEE Inter Society Conference on Thermal Phenomena, pages 626-634, 2002.
34. Despatch Industries. http://www.despatch.com/pdfs/PBC.pdf.
35. V. Szekely. "Thermal monitoring of microelectronic structures". Microelectronic Journal, 25(3):157-170, 1994.
36. B. Lian, T. Dishongh, D. Pullen, H. Yan, and J. Chen. "Flow network modeling for improving flow distribution of microelectronics burn-in oven". IEEE Inter Society Conference on Thermal Phenomena, pages 78-81, 2000.
37. D. Gardell. "Temperature control during test and burn-in". IEEE Inter Society Conference on Thermal Phenomena, pages 635-643, 2002.
38. A. Poppe, G. Farkas, M. Rencz, Z. Benedek, L. Pohl, V. Szekely, K. Torki, S. Mir, and B. Courtois. "Design issues of a multi-functional intelligent thermal test die". Proceedings of IEEE SEMI-THERM Symposium, pages 50-56, 2001.
39. http://www.spraycooling.com.
40. T. Kim and W. Kuo. "Modeling manufacturing yield and reliability". IEEE Transaction on Semiconductor Manufacturing, Vol. 12, No. 4, pages 485-492, 1999.

41. T.S. Barnett, A.D. Singh, M. Grady, and K.G. Purdy. "Redundancy implications for product reliability: Experimental verification of an integrated yield-reliability model". Proceeding of International Test Conference, pages 693-699, 2002.

42. J. Van der Pol, F. Kuper, and E. Ooms. "relation between yield and reliability of integrated circuits and application to failure rate assessment and reduction in the one digit fit and ppm reliability era". Microelectronics and Reliability, Vol. 36, No. 11/12, pages 1603-1610, 1996.

43. T.S. Barnett and A.D. Singh. "Relating yield models to burn-in fall-out in time". Proceeding of International Test Conference, pages 77-84, 2003.

44. T. Kim, W. Kuo, and W.K. Chien. "Burn-in effect on yield". IEEE Transaction on Electronics Packaging Manufacturing, Vol. 23, No. 4, pages 293-299, 2000.

45. N.F. Dean and A. Gupta. "Characterization of a thermal interface material for burn-in application". Proceedings of IEEE Thermal and Thermo-mechanical Phenomena in Electronic Systems, pages 36-41, 2000.

46. W. Wondrak. "Physical limits and lifetime limitations of semiconductor devices at high temperature". Microelectronics Reliability, Vol. 39, No. 6-7, pages 1113-1120, 1999.

47. International Technology Roadmap for Semiconductors (ITRS). http://public.itrs.net/.

48. H. Sanchez, B. Kuttanna, T. Olson, M. Alexander, G. Gerosa, R. Philip, and J. Alvarez. "Thermal management system for high performance PowerPC microprocessors". Proceedings of IEEE COMPCON, pages 325-330, 1997.

49. P. Reed, M. Alexander, J. Alvarez, M. Brauer, C.C. Chao, C. Croxton, L. Eisen, T. Le, T. Ngo, C. Nicoletta, H. Sanchez, S. Taylor, N. Vanderschaaf, and G. Gerosa. "A 250-MHz 5-W PowerPC microprocessor with on-chip L2 cash controller". IEEE Journal of Solid-State Circuits, Vol. 32, No. 11, pages 1635-1649, 1997.

50. A. Vassighi, O. Semenov, and M. Sachdev. "Impact of power dissipation on burn-in test environment for sub-micron technologies". Proceedings of IEEE International Workshop on Yield Optimization and Test, pages 1-5, 2001.

Chapter 4

THERMAL AND ELECTROTHERMAL MODELING

Abstract: In this chapter we begin with an overview of thermal modeling of high performance microprocessors with emphasis on the necessity of these models to ensure a reliable system. Later in this chapter we review the concept of the electro-thermal modeling and describe several approaches to electro-thermal modeling at architecture and circuit levels.

Key words: Thermal Modeling, Electrothermal Modeling,

1. OBJECTIVES OF THERMAL ANALYSIS

In recent years, power density in microprocessors and other high performance circuits has doubled every two to three years and this rate is expected to increase in coming generations as feature sizes, threshold voltages and frequencies scale faster than operating voltages. The energy consumed by a high performance chips including microprocessors is converted to the heat. As a consequence, realization of an electrical network inherently includes corresponding realization of a thermal network as well.

All physical components (as piece of mass) act as heat storage capacitances, known as *thermal capacitances* (C_{th}), and the heat dissipated by them is conducted towards the ambient by means of heat transfer through *thermal resistances* (R_{th}).

These two electrical and thermal networks are not independent and are coupled to each other. The dissipating elements of the electrical network are heat sources of the thermal network. On the other, the temperature values in the thermal network in different nodes also affect the electrical parameters of the components in a significant way. Most of the device electrical parameters have strong temperature dependence. This means that the behavior of the *IC* can not be obtained from the electrical network alone but will be influenced by the thermal network and thermal couplings amongst its heat sources. The thermal coupling can influence the function of the circuit and in extreme cases it can even cause thermal runaway and destruction of the chip.

Today, circuit simulators are used as standard tools in the design and optimization of electronic circuits. However, these tools have limited capabilities incorporating temperature as a dynamic variable in simulations. Typically, temperature is specified as a static, global variable. Owing to heterogeneous nature of contemporary designs, chips often have thermal gradients which can not be included in circuit simulation environment. In electronic systems, the temperature is one of the critical parameters owing to its strong impact on transistor parameters.

Within the safe operating conditions, temperature fluctuations as well as gradients strongly affect the lifetime of the semiconductor components. Several aging mechanisms such as gate oxide breakdown, electro-migration are thermally accelerated. Hence, a higher localized junction temperature will cause accelerated localized failures. On the other hand, the temperature fluctuations create mechanical stresses. A VLSI is composed of materials with different thermal coefficients of expansion. As a consequence, temperature fluctuations result in mechanical stresses. Each change in temperature causes mechanical stress in the component which, in particular, affects solder and bond connections. Here it is not the absolute temperature which causes the problem but the temperature cycling. As a rule of thumb it can be assumed that the aging of a component is proportional to the fourth power of the temperature deviation [1].

Other than the mechanical stress, temperature variation also causes electrical stresses in the chip. The *ON*-resistance of a *MOSFET* increases with temperature increase and causes conduction losses in a transistor. The conduction losses are almost doubled when temperature increases from 25°C to 150°C [2]. On the other hand the threshold voltage of a *MOSFET* decreases with increasing temperature which reduces the signal-to-noise margin at the control node. Reduced threshold voltage also increases the leakage current of the transistor. Ignoring these effects can lead to an undesired even catastrophic *turn-on* of the transistor when it should be off. In addition to these issues, circuit designer must investigate the circuit performance under wider, fluctuating temperature range. For example, in relay protection circuits, the avalanche energy of power *MOSFETs* is specified in such a way that any load current pulse pattern below the rated current is allowed as long as the peak junction temperature does not exceed the maximum specified value.

Table 4-1. Temperature dependency of important *Si-MOSFET* parameters [90].

Parameter	Temperature Dependence	Affected Property
Thermal Conductivity (K)	$=T^{1.6}$	Self Heating
Built-in Potential (V_{bi})	$\dfrac{KT}{q}\ln\left(\dfrac{N_A N_D}{n_i T^2}\right)$	+20% per 100K
Threshold Voltage (V_{TH})	$2Si_B(T)+$ $(4\varepsilon_{Si}qN_A Si_B(T)/C_i)^{0.5}$	-0.8 mV/K
pn Junction Reverse Current	$a\times n_i^2(T)+$ $b+n_i(T)/T_{SC}$	10^2 to 10^4 per 100K

While performing the thermal analysis, it is important to know the physical limits of maximum operating temperature on semiconductor substrate. The maximum operating temperatures for semiconductor devices can be estimated from semiconductor intrinsic carrier density, which depends on the band-gap of the material. When the intrinsic carrier density reaches the doping level of the active region of a device, at that instance, the electrical parameters change drastically. The highest operating junction temperature for standard silicon technology is about 200°C [3]. The influence of temperature on some important *MOSFET* parameters is summarized in Table 4.1. Several practical considerations limit the junction temperature to a much lower value. A limit of 150°C for junction temperature is often used for VLSI *ICs* [4]. The peak junction temperature of a PowerPC microprocessor implemented in a 350 nm *CMOS* technology was reported to be approximately 90°C-100°C at an operating speed of 200-250 MHz [5][6].

2.   THERMAL NETWORK MODELING

In this section we will describe the modeling of the thermal network which coexists with every electrical network in the electronic systems. The heat propagation in a system can take place in three different ways, convection, radiation, and conduction. In electronic components heat usually propagates through the conduction which often assumed to be one dimensional, for the sake of simplicity. Eq. (4.1) describes one-dimensional heat conduction in a homogeneous isotropic material.

$$\frac{\partial^2 T}{\partial x^2} = \frac{c.\rho}{\lambda_{th}} \cdot \frac{\partial T}{\partial t} \qquad (4.1)$$

In Eq. (4.1), λ_{th} is the heat conductance, c is the thermal capacitance and ρ is the density of the material. T is the temperature and x is the direction of the heat flow in the material. The electrical model which is equivalent to this thermal model is the model for a transmission line which can be described as Eq. (4.2).

$$\frac{\partial^2 U}{\partial x^2} = CL\frac{\partial^2 U}{\partial t^2} + (CR + GL)\frac{\partial U}{\partial t} + GRU \qquad (4.2)$$

In this equation, C is the capacitance per unit area, R is the resistance per unit area, L is the inductance per unit area and G is the transverse conductance per unit area. Eq. (4.2) describes all the property of a wave in a transmission line. Considering the fact that in the thermal network where heat is transmitted in a solid media, there is no direct comparison for the electrical conductance (element cannot cool itself) or in other words $L=0$ and $G=0$, the Eq. (4.2) can be reduced to Eq. (4.3).

$$\frac{\partial^2 U}{\partial x^2} = CR\frac{\partial U}{\partial t} \qquad (4.3)$$

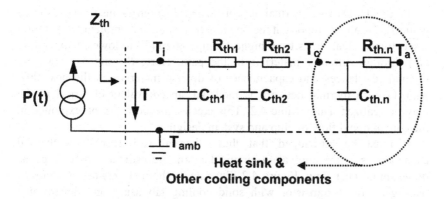

Figure 4-1. Principle circuit diagram of a model with interactive coupling of electrical and thermal component description.

Equation (4.3) has the same structure as Eq. (4.1) which is for heat conduction and as Kirchhoff stated in 1845, "Two different forms of energy behave identically when the basic differential equations which describe them have the same form and the initial and boundary conditions are identical" [7]; therefore, heat conduction process can be modeled by a transmission line equivalent circuit diagram which, as shown in Figure 4.1, consists of *R/C* elements only. Table 4.2 depicts the equivalent parameters between electrical and thermal domains.

Table 4-2. Equivalent thermal and Electrical variables.

Thermal		Electrical	
Temperature	T in K	Voltage	U in V
Heat Flow	P in W	Current	I in A
Thermal Resistance	R_{th} in K/W	Resistance	R in V/A
Thermal Capacitance	C_{th} in Ws/K	capacitance	C in As/V

Figure 4.1 shows a thermal network where $P(t)$ represents the heat source while T_j, T_C, and T_a represent the on chip, heat sink, and ambient temperature respectively. Heat flows from highest temperature (T_j) to lowest temperature (T_a) through different materials in the path. Here, R_{th} and C_{th} represent thermal resistances and capacitances of diverse materials in the heat flow path. Once a thermal network is modeled, an equivalent electrical model may be realized using Table 4.2. This electrical model can be exploited in the conventional flow of chip and system design.

It must be mentioned that these thermal resistances and thermal capacitances in the equivalent circuit diagram are estimated assuming one dimensional heat flow. Figure 4.2 illustrates a thermal network model of a packaged power transistor with solid cooling tab using one dimensional thermal network modeling approach. For complicated heat flow conditions, a three dimensional thermal equivalent network may be devised [2].

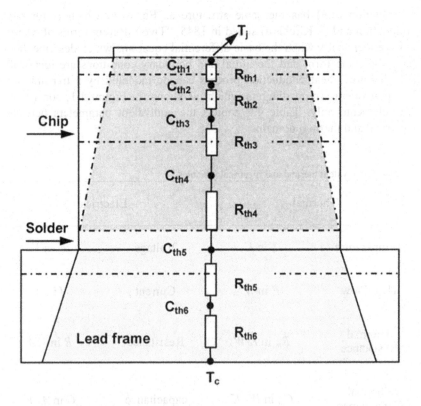

Figure 4-2. The thermal equivalent elements can be derived directly from the physical structure.

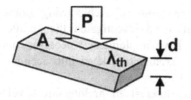

Figure 4-3. Heat flowing through a material with thickness of d and area of A and conductivity of λ_{th}.

Using Figure 4.3, the thermal resistance and the thermal capacitance of the material can be determined by Eq. (4.4) and Eq. (4.5), respectively [2].

$$R_{th} = \frac{d}{\lambda_{th} \cdot A} \tag{4.4}$$

$$C_{th} = c \cdot \rho \cdot d \cdot A \tag{4.5}$$

In these equations A is the area and d is the thickness of the material. Using the model in Figure 4.2, the temperature at the beginning of the component (T_{j1}) can be determined by the power passing through that component and the temperature at the end of the component (T_{j2}).

$$T_{j1} = T_{j2} + Z_{th1} \times P \tag{4.6}$$

The same concept may be applied to the whole chip and by using Figure 4.1, the junction temperature of the chip (T_j) can be described as:

$$T_j = T_a + Z_{th} \times P \tag{4.7}$$

Where T_a is ambient temperature, P is the dissipated power in the chip, and Z_{th} is the total thermal impedance of the complete network.

3. ARCHITECTURAL LEVEL ELECTROTHERMAL MODELING

Circuit design techniques optimize heat dissipation for a particular circuit style and influence locally, while architecture level design techniques change the behavior of large segments of a chip using global knowledge of the design. Skadron et. al. have investigated the architecture level electrothermal modeling and have implemented necessary tools to take thermal behavior of the chip at architecture level into account. In this section we briefly review their work.

There is a growing interest in architecture level solutions, such as dynamic voltage scaling and fetch throttling to address the thermal stress related issues [8-10]. In order to accurately estimate the power and performance of a chip, designers and architects need a way to model the temperature at any level of design abstraction. In this context, accurate thermal modeling plays an important role since performance and power are strongly depend on the thermal map of a specific implementation or architecture. Architectural level electrothermal modeling is a two step procedure; (i) floor plan extraction, and (ii) thermal *RC* modeling [11].

3.1   Floor Plan Extraction

Figure 4.4 shows a sample floor plan of Alpha 21264 processor [12]. To derive a high level floor plan often several constraints such as adjacencies, architectural, functional, and other user specified, must be taken into account.

In order to implement the architecture level floor plan, the user maps the floor plan as a set of blocks as illustrated in Figure 4.4. Each block is subdivided into a matrix of sub-blocks. Each sub-block corresponds to a set of functional units such as memory row or column. Sub-blocks are fit into the block by adjusting their widths and sub-block aspect ratio is assumed to be variable. Such a flexible arrangement allows the user to completely specify a floor plan using vertical and horizontal adjacency matrices.

3.2   Lumped Thermal RC Network

The next step in thermal modeling of a chip is to derive an equivalent circuit to model dynamic heat flow in the chip and to compute the values of the thermal resistances (R_{th}) and capacitances (C_{th}). This is accomplished using the floor plan with areas and thermal resistance and thermal capacitance values for the package. The dissipated power in each block is received by the thermal model in every time step. Then the average temperature of each of these blocks is calculated at the end of that time step. The circuit model of a simple chip with three blocks is shown in Figure 4.5.

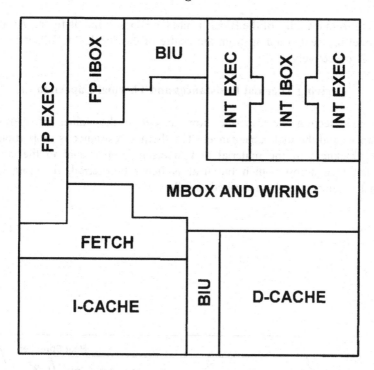

Figure 4-4. Normalized floor plan for the Alpha 1264 processor.

Figure 4.5 shows thermal resistances and thermal capacitances from each of the blocks to all its neighbors (not all of them are shown). This figure also shows $R_{th}C_{th}$ pairs from the centre of each block to common nodes which represent the flow of heat from each block into the package. The electrothermal modeling tool must compute these thermal resistance and the thermal capacitance values. This captures spatial non-uniformity in temperature in the package. Finally, a fixed R_{th} and C_{th} are used to model the heat sink. These values of R_{th} and C_{th} depend on the physical parameters and on the geometry of blocks and of the die. Although in this model the interface materials between the die, spreader, and heat sink are neglected, but they can be modeled the same fashion as other packaging materials.

As it can be seen in Figure 4.5, the *RC* model consists of a vertical model and a lateral model for the die, spreader, and heat sink layers. The vertical composite $R_{th}C_{th}$ models the heat flow from one layer to the next, starting from the die, going through the package and finally into the air. The lateral composite $R_{th}C_{th}$ models the heat diffusion (i.e., coupling) between neighboring blocks within a layer, and from the edge of one layer into the periphery of the next area. For example, in Figure 4.5, *R1* models the heat coupling from the edge of Block 1 into the spreader and *R2* models the heat coupling from the edge of the Block 1 into the Block 2. In this figure, *R3* and *R4* model the heat coupling from the centre of the Block 3 to Block 1 and Block 2 respectively.

3.2.1   Deriving Thermal Resistances and Thermal Capacitances

In this section we describe how to derive the values of thermal resistances and thermal capacitances. The thermal resistance is proportional to the thickness of the material and inversely proportional to the cross sectional area across which the heat is being transferred, therefore the thermal resistance can be described as follow:

$$R = \frac{t}{k \times A} \tag{4.8}$$

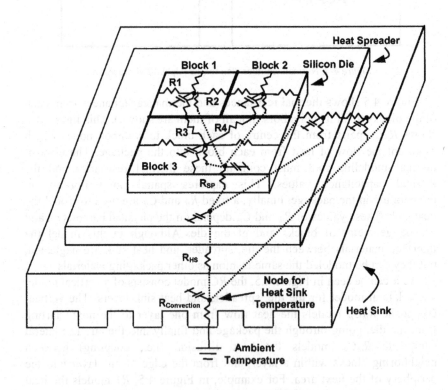

Figure 4-5. Sample circuit model for simple floor plan of three blocks.

In Eq.(4.8), t is the thickness of the material, A is the area of the material and k is the thermal conductivity of the material per unit volume. k for silicon, is $100W/m^3.K$ and for copper is $400W/m^3.K$ at $85°C$. On the other hand the thermal capacitance is proportional to both thickness and area and can be described as follow:

$$C = c \times t \times A \qquad (4.9)$$

Where, c is the thermal capacitance per unit volume. The value of c for silicon is $1.75x10^6$ $J/m^3.K$ and for copper is $3.55x10^6$ $J/m^3.K$. In addition, lateral resistances must take the spreading resistance between blocks of different aspect ratios into account, and the vertical resistance of the heat sink must take the constriction resistance from the heat sink base to the fins into account [13]. Spreading resistance accounts for the increased heat flow from a small area to a large one and vice versa for constriction resistance [14]. It must be noted that the capacitance values may need to be adjusted by an empirical fitting parameter due to the fact that the bottom surface of the chip is not isothermal, whereas the described model treats the bottom surface as isothermal to reduce model complexity. However, the true isothermal surface lies in the heat sink, which means the equivalent thermal mass that determines the rate of on-die heating is larger than the die thickness and makes the effective thermal capacitance larger than what Eq. (4.9) yields.

4. ELECTROTHERMAL MODELING AT LOGIC LEVEL

In large digital designs, transistor level electro-thermal simulation is very expensive in terms of computing resource. Moreover, the fine detailed information may not be necessary. The thermal time constants associated with thermal network of a chip are larger by many orders of magnitude than the typical clock cycle of digital systems. The relaxation type simulation schemes can be applied in such cases to reduce the computational complexity. In addition, in case of logic gates, there is no need for highly accurate circuit level thermal simulation. Cheng et. al therefore suggested that simpler models and simulators are sufficient [15, 16]. In their work they applied the *RWQ* (region-wise quadratic) *MOS* transistor modeling technique in order to speed up their simulator.

Szekely et. al. [17] suggested that relaxation type simulation is still too detailed with respect to the final design objectives and the relaxation method should be utilized for coupling a logic simulator (using logic models of gates, flip-flops and other standard digital building blocks) to fast thermal simulator. They implemented a logic level thermal simulator assuming the gate level simulation and standard cell design style. Their modeling method involves the following steps.

Identifying the thermal behavior of the logic gates, flip-flops and other standard cells. This means that for a logic block, the temperature dependence of their timing parameters must be extracted. Furthermore, the switching power and leakage power including their temperature dependence parameter must be obtained. This can be achieved by extracting the electrical net-list of the each standard cell or logic block and performing electrical simulations on the block while sweeping the temperature in the desired range (from 0°C to 100°C).

Once the thermal characteristics of standard cells or logic blocks are extracted, temperature dependent simulation models should be constructed for the purpose of logic simulation, where each logic model has the temperature as parameter. Using the transistor level simulations the temperature dependent logic models of the standard cells should be constructed only once and should be included in the standard cell library.

Since the event count annotated to each logic gate is needed to calculate the power dissipation of the gate, the applied logic simulator has to be modified such that it evaluates these event counts for each logic gate during time step Δt. Then multiplying this number by single event power dissipation of the logic gate, results in the total power dissipation of the logic gate during a given time interval Δt. After time interval of Δt this dissipation distribution of the circuit will be passed to the thermal simulator in the next step.

In this step, the thermal simulator has to calculate the temperature distribution of the chip. The thermal simulator uses the power distribution which is passed to it from the logic simulator after time interval Δt and also the layout of the chip together to calculate the temperature distribution in the chip. The updated gate temperatures will be sent back to logic simulator to update the temperature dependant parameters like timing and power dissipation parameters, using their temperature dependant models in the logic simulator. Once these models and their parameters are updated, the logic simulation has to be performed for another time interval Δt. This procedure will be continued till final simulation time is exceeded.

5.   ELECTROTHERMAL MODELING AT CIRCUIT LEVEL

Szekely et. al. has carried an extensive research regarding the electrothermal modeling of the VLSI chips at circuit level. In this section we review some of their work [17-22]. As it was mentioned before, the IC is composed of two coupled networks: an electrical network consists of electrical circuit which accomplishes the design objective, and a thermal network which represents the thermal environment of the silicon chip. These two networks are coupled through the power dissipation of the electrical circuit which acts as the heat source. The heat flows from the die to the ambient. This coupling is reinforced through temperature dependency of the components of the electrical circuit and semiconductor device parameters.

The mutual dependence of subsystems which describe these networks can be formulated by the corresponding state equations [17]. For electrical side the formulation is as follows:

$$0 = I_m(V_i, T_j) \tag{4.10}$$

And for the thermal part is:

$$0 = P_n(V_i, T_j) \tag{4.11}$$

In Eq. (4.10) and Eq. (4.11), V_i is the vector of the electrical state variables like nodal voltages, T_j is the vector of the state variables of the thermal subsystem like device temperatures, I_m is the net current entering node m and P_n is the heat-flux balance of the n-th device's thermal node. The I_m and P_n functions are nonlinear in both arguments. The coupled electrothermal system is characterized by the common state vector $[V_i, T_j]$, that we seek during the solution process. For solving these equations either the relaxation method or the method of simultaneous iteration can be applied.

In simultaneous iterative simulations, Eq. (4.10) and Eq. (4.11) are considered to form one, single system of equations for the concatenated, common, state vector. The iterative solution takes place simultaneously for both voltages and temperatures. It is characteristic for this method that during the iterative solution process, such as in case of the Newton-Raphson algorithm, all elements of the Jacobian matrix are calculated simultaneously. Different classical implementations of this method can be found in the literature [22-25].

In case of the relaxation method, for the electrical and for the thermal parts two separate solution algorithms are used. These solutions are coupled such that one simulator uses the updated results of the other simulation tool in an iteration loop. Examples of electro-thermal simulation systems based on this principle are presented by various authors [26-28]. The drawback of this method is that the $\partial I_m/\partial T_j$ and $\partial P_n/\partial V_i$ elements of the Jacobian matrix of the coupled system are not generated at all. That is why such a simulation system is unable to deliver immediate results regarding the small signal, frequency domain behavior. Therefore, for transistor-level electro-thermal simulation we describe the method of simultaneous iteration.

5.1   Modeling Issues

- For an accurate, self-consistent electro-thermal simulation performed on a circuit, following must be addressed:
- A thermal port must be added to the semiconductor device models of the applied circuit which includes extra device equations in order to account for basic electro-thermal effects (dissipation, temperature dependence of the physical parameters).
- Temperature sensitive elements other than the semiconductor devices such as *Si-Al* contacts must be included in the models.
- An accurate and efficient lumped element thermal *RC* model of the thermal subsystem must be generated.

5.1.1   Advances in the Thermal Net-list Extraction

One of the important issues in electrothermal simulation with simultaneous iteration method is how to identify the lumped element equivalent of the distributed thermal *RC* system corresponding to the silicon chip. Figure 4.6 shows that how the thermal net-list is extracted from the electrical network.

The thermal part of an electro-thermal system is composed of an electrical/thermal model with N thermal ports in electrical network and a thermal N-port. The relevant devices of the electrical circuit are connected to the ports of the thermal N-port (Figure 4.6). In other words, this thermal N-port terminates the thermal ports of the electro-thermal device models (semiconductor device models, model of the Si-Al contacts) of the electrical N-port. The thermal N-port can be described as follows:

$$T_i = f_i(P_j), i = 1, \cdots n, j = 1 \cdots n \qquad (4.12)$$

Where T_i is the temperature response at the i-th port and P_i is the power consumption at the j-th port of the thermal network.

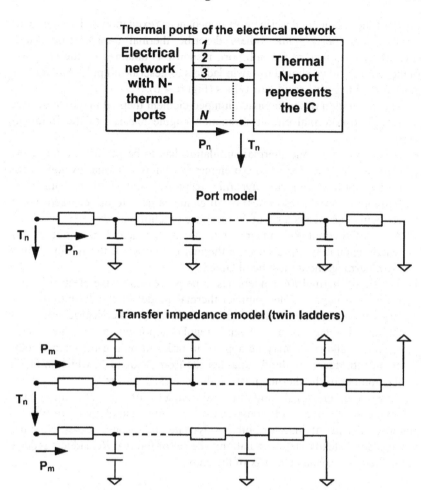

Figure 4-6. The model of coupled electro-thermal system realized by an *IC* and the lumped element *RC* equivalent of the *IC* chip describing thermal behavior.

To analyze a given layout arrangement of the *IC*, the f_i function matrix must be obtained by means of a thermal simulator. Then, using f_i, an equivalent circuit must be generated that properly describes the thermal subsystem. For the steady-state case the network consist of thermal resistors. In other words, for a *dc* electro-thermal analysis, the thermal behavior of the chip and its thermal surrounding can be accurately described by the thermal resistance matrix. For dynamic or *ac* analysis, the equivalent circuit of the thermal *N*-port is a set of lumped element *RC* ladders, shown in Figure 4.6.

Some of these ladders describe port impedances (excitation and response are on the same port), some others describe thermal transfer impedances (excitation is applied to the mth port; response is obtained on the nth port). In the latter case we should use twin ladders to describe all poles having both positive and negative residuum values (Figure 4.6).

The generation of the compact, lumped element thermal model to be used by the electro-thermal circuit simulation engine consists of the following steps:

- A series of dynamic thermal simulation has to be performed. On every element (temperature sensitive element) a known thermal excitation has to be applied and the dynamic responses (either time domain or frequency domain responses) of all elements have to be recorded. For N elements this means a set of N^2 response functions.
- A set of thermal RC ladders (with K stages where K is about 10-16), which accurately represents the thermal properties of the IC for a given layout arrangement must be obtained.
- The set of thermal RC ladders has to be processed by the electro-thermal simulation engine. The compact thermal model of the IC introduces a large amount of excess nodes to be considered by the electric solver. By taking advantage of the special, regular topology, a dedicated node-reduction algorithm may be applied in order to reduce the excess node-count introduced by the thermal net-list from $N^2 xK$ to N, which in fact is the number of the thermal ports of the electrical net-list.

Since the electrical and thermal models of the IC are treated simultaneously, the self-consistency of the simulation results is automatically maintained for both steady-state and dynamic cases. Note that in the case of steady-state simulations, the set of thermal RC ladders reduces to the thermal resistance matrix of the chip.

5.1.2   Si-Al contacts

Due to the Seebeck-effect the *Si-Al* contacts also serve as coupling elements between the electrical and thermal parts of the electro-thermal system. They can be considered as thermo-elements, modeled by a temperature controlled voltage source with a thermo-electrical transconductance of $S \approx 0.2\text{-}1mV/^{\circ}C$. Besides the circuit simulation model, the consideration of the *Si-Al* contacts involves several other problems. The most important one is that the usual circuit extractors are not able to detect any electrical contact in the layout; layout features connected by *Si-Al* contacts are considered to be equi-potential, thus they form one single node in the electrical net-list. This problem can be solved by adding new detection and net-list generation rules to the rules data base of the net-list extractor. The operation of the layout extraction can be interactively controlled by the user inside a design environment: certain sub-regions of the layout may be marked where the *Si-Al* contact should be detected as temperature controlled voltage sources. For the contacts the corresponding shapes also appear in the layout description forwarded to the thermal solver.

6. ELECTROTHERMAL MODELING AT DEVICE LEVEL

Different methods have been developed to characterize the electrothermal behavior of the electronic devices. These methods include the numerical methods, the Fourier series expansion, and the elemental heat source approach. Numerical methods such as finite elements [29, 30], boundary elements [31], finite differences and thermal networks [32-34, 28] account for nonlinear thermal effects, but they are computationally intensive. Moreover, in these methods, the heat sources are represented as point source and therefore the geometrical information of the heat source are lost which compromises the accuracy [32].

The Fourier series expansion technique consumes less *CPU* time and can handle multiple layer structures [35-38]. In this technique the number of terms that must be taken into account is proportional to the ratio of the chip to heat source size. Therefore, for structures with large chip to heat source ratios, the *CPU* time becomes very long and makes this technique unsuitable for complex layout geometries. This problem can be solved partially by using double Fourier transform technique [39], but due to numerical integration procedure in this technique, the *CPU* time is still high.

The third approach for the thermal analysis of electronic devices evaluates the contribution to the temperature field produced by an elemental heat source. Then the contribution of all these elements is superimposed to calculate the total temperature distribution. Rinaldi [40] proposed a closed form analytical approach to the heat flow problem in integrated circuits. His technique was an extension to the technique proposed by Smith et. al. [41], which was able to calculate the surface temperature distribution by placing a thin rectangular heat source on the top surface of a structure with finite lateral dimensions and infinite thickness. Rinaldi extended this technique in several aspects. First, the temperature field was computed for the entire chip. Then, the boundary conditions in the top and the bottom were taken into account using the method of images. Therefore, the thickness of the chip was included into the analysis. Since in many devices, the region where the heat is generated can not be approximated to an infinitely thin rectangle, Rinaldi's solution also considers arbitrarily located embedded heat sources.

In comparison with the Fourier series and the Fourier transform techniques, Rinaldi's technique reduces the *CPU* time considerably. Using this approach, a simple and accurate expression (Eq. 4.13) for the thermal resistance of an integrated device has been obtained which takes the exact location and shape of the heat source and a finite substrate thickness into account.

$$R_{ja} = \frac{1}{2\pi k}\left[\frac{1}{L}\ln\left(\frac{L+\left(W^2+L^2\right)^{0.5}}{-L+\left(W^2+L^2\right)^{0.5}}\right) + \frac{1}{W}\ln\left(\frac{W+\left(W^2+L^2\right)^{0.5}}{-W+\left(W^2+L^2\right)^{0.5}}\right)\right]$$

(4.13)

7.   STATIC ELECTROTHERMAL MODELING: A CASE STUDY

Static electro-thermal modeling is used to study the behavior of the chip under steady state conditions; while dynamic electro-thermal modeling follows instantaneous changes in the junction temperature of the chip. These conditions, static or dynamic, include inputs to the chip, packaging of the chip, and cooling solution that are implemented to remove the heat from the chip. This kind of modeling is suitable for studying the area, power and performance tradeoffs when different kind of packaging and cooling solutions are used for removing the heat from the chip.

To account for the change in the junction temperature, an electro-thermal analysis tool has been developed to self-consistently compute the junction temperature, power, and operating frequency of a microprocessor owing to its high power consumption. However, the tool can be utilized for any complex VLSI chip. Figure 4.7 shows the framework of this electro-thermal optimization tool [42, 43]. Starting with an initial assumption of junction temperature ($T_{j-initial}$), the electro-thermal analysis tool first computes frequency and power. At this point, data from simulations and measurements for I_{on} and I_{off} at the given V_{DD} and given junction temperature ($T_{j-initial}$ for the first iteration) are incorporated into power and frequency calculations. Other parameters are extracted from process files. In the next step, based on the package and cooling system characteristics, the new junction temperature (T_j) is computed, and the new T_j is the starting point for the power and frequency calculations in the following iterations. These iterations continue until junction temperature computed in consecutive steps converges to a steady-state temperature. If the convergence is not achieved, it indicates the thermal runaway [44]. On the other hand, if the iterations converge, the final self-consistent junction temperature is obtained. The tool also produces corresponding values for frequency, chip switching and leakage power components, active cooling system power, and the die area that are consistent with the final temperature.

The tool requires following input parameters:

- Thermal resistance of the packaging and cooling system.
- Coefficient of performance (*COP*) for active cooling (defined as the ratio of the chip power to the power consumed by the cooling system).

Chip design and process technology characteristics. Design and technology parameters are illustrated in Figure 4.8.

An extensive research is conducted to follow each step in this algorithm to keep the tool physically-based. The frequency calculations mimic microprocessor frequency limitations by considering critical path delay and the role of interconnect delay. Transistor parameters including I_{on}, I_{off}, C_j, and C_{gate} and interconnect parameters including C_{int} and R_{int} are extracted, measured, or computed. Subsequently, transistor and interconnect parameters, together with the other circuit parameters such as supply voltage, body bias voltage, number of buffers used in long interconnect lines, and logic depth in critical paths, are utilized to calculate power and operational frequency. The critical path logic depth is used to transition from the transistor to the microprocessor frequency calculation. Figure 4.8 illustrates the important physical parameters which are used to develop the electro-thermal analysis tool. In this figure, each row corresponding to Input column (frequency, power and temperature calculations), depicts all the relevant parameters. Each of these parameters, their changes with respect to other parameters, and their impact on the simulation results has been investigated separately. Their values have been extracted from process files, measurements, or physically based calculations.

After incorporating these parameters into the tool, the tool has been calibrated to actual microprocessor measurements. Figure 4.9 shows the sort level measured data and simulated data for $V_{DD}=1.1V$ to $V_{DD}=1.7V$ and $T_j=25^\circ C$. The calibration has been performed by modifying the critical parameters in architecture and circuit level. These parameters include number of stages in critical path and the activity factor of the chip. Other measurements like total leakage power, total dynamic power, and cooling power have been used to confirm the simulation results with respect to measurement results.

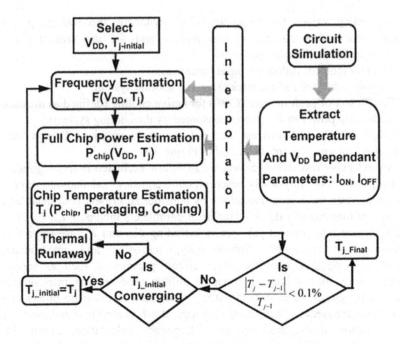

Figure 4-7. The framework of the electro-thermal analysis tool.

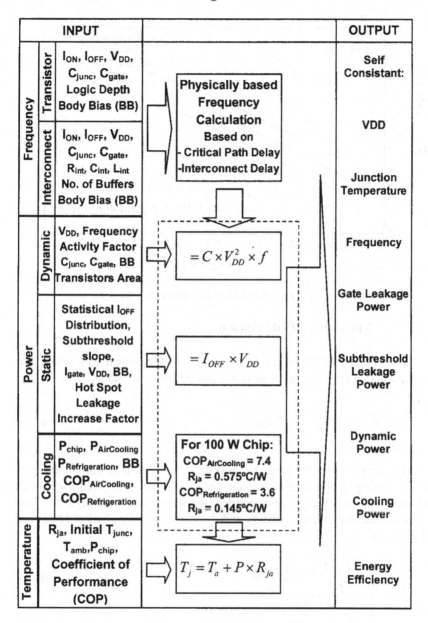

Figure 4-8. Algorithm of the self consistent physically based electro-thermal modeling approach.

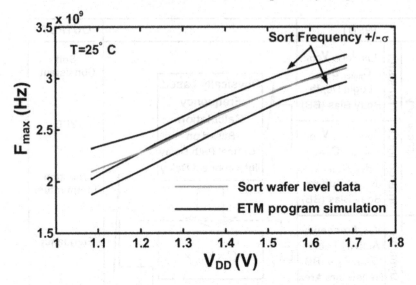

Figure 4-9. Sort level data and measured data for calibration of the electro-thermal tool.

7.1   Frequency Calculation

The operational frequency is the inverse of the critical path delay. The critical path delay is the average of the critical path rise and fall delays.

$$\tau = \frac{1}{f} = \frac{\tau_{charge} + \tau_{discharge}}{2} \tag{4.14}$$

The critical path delay can be calculated based on total current charging and discharging the load capacitance as shown in Figure 4.10 and can be defined as the average time needed to charge and discharge the load capacitance [46]. Time to charge and discharge are given by Eq. (4.15) and Eq. (4.16).

$$\tau_{charge} = \frac{C \times V_{DD}}{I_{charge}} \times \frac{1 + f_{int}}{n} \tag{4.15}$$

$$\tau_{discharge} = \frac{C \times V_{DD}}{I_{discharge}} \times \frac{1 + f_{int}}{n} \tag{4.16}$$

During the charge, the *PMOS* transistor is on and the *NMOS* transistor is off and as a result, most of the current from *PMOS* transistor charges the load capacitance. However, a part of the *PMOS* current passes through the off *NMOS* transistor (leakage). Therefore, the net charging current can be expressed as follows:

$$I_{charge} = I_{on-P} \cdot W_P - I_{off-N} \cdot W_N \tag{4.17}$$

During the discharge, the situation is complementary. Therefore, the net discharging current can be expressed as follows:

$$I_{discharge} = I_{on-N} \cdot W_N - I_{off-P} \cdot W_P \tag{4.18}$$

Using Eq. (4.14) to Eq. (4.18), the frequency can be calculated as follows:

$$f = \frac{2 \cdot n}{(1 + f_{int}) \cdot C \cdot V_{DD}} \times$$

$$\frac{(I_{on-N} \cdot W_N - I_{off-P} \cdot W_P) \cdot (I_{on-P} \cdot W_P - I_{off-N} \cdot W_N)}{(I_{on-N} \cdot W_N - I_{off-P} \cdot W_P) + (I_{on-P} \cdot W_P - I_{off-N} \cdot W_N)} \tag{4.19}$$

In Eq. (4.15) and Eq. (4.16), n is the logic depth in the critical path and f_{int} is the fraction of the capacitance contributed by interconnects. I_{on-N} and I_{on-P} are the on-currents of *NMOS* and *PMOS* transistors, respectively, where I_{off-N} and I_{off-P} are the off-currents of these transistors. In Eq. (4.17) to Eq. (4.19), C is the total capacitance in one stage and W_N and W_P are the total widths of *NMOS* and *PMOS* devices respectively. Note that the *on*-current and *off*-current of transistors can be obtained from circuit simulations. In Figure 4.10, C_{load} is the load capacitance.

7.2 Full Chip Power Estimation

For a microprocessor, the total power can be expressed as:

$$P_{total} = P_{logic} + P_{memory} + P_{I/O} \tag{4.20}$$

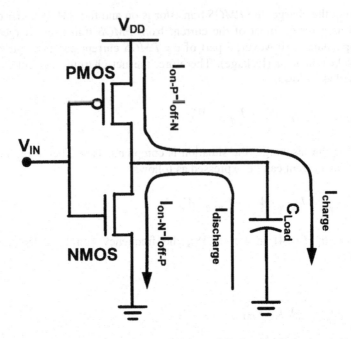

Figure 4-10. The total currents that are charging and discharging the load capacitance.

The logic in the microprocessor can be sub-divided into various functional blocks such as the data path, register files, control, etc. The total width of devices in each block as well as in memory (cache) and in *I/O* blocks is extracted from the actual designs. For each of these blocks, the leakage power and dynamic power based on the maximum achievable frequency can be calculated.

For power calculations, switching power, leakage power and cooling power must be considered. The short circuit power can be ignored for simplicity. Dynamic switching power must be computed according to the appropriate microprocessor activity factor, chip supply voltage, chip switching capacitance, body bias, and the chip area. Static leakage power must also include gate leakage. The chip leakage is derived based on statistical transistor leakage distributions [45]. The role of hot spots on the chips must be considered for leakage and maximum operating frequency of the chip by giving weight to different blocks on the chip based on the temperature distribution which has been derived from experimental data. Cooling power is computed based on chip power and the *COP* and thermal resistance of different cooling solutions.

7.2.1 Dynamic Power Estimation

The dynamic power or switching power of a chip can be expressed in terms of its blocks as described in Eq. (4.21) [45]:

$$\sum_{i=1}^{N} P_{i-switching} = \sum_{i=1}^{N} \alpha_i \cdot C_i \cdot V_{DD}^2 \cdot f(V_{DD}, T_j) \tag{4.21}$$

Where N is the total number of blocks in the chip and α_i is the activity factor and C_i is the total capacitance of a block. V_{DD} is the specified operating voltage, f is the chip operating frequency and T_j is the chip junction temperature.

7.2.2 Capacitance Calculation for Frequency and Power Estimation

The critical path of a microprocessor can be represented with a cascade of same sized inverters. One such inverter is depicted in Figure 4.10. In order to estimate realistic power and frequency of the critical path, and hence, the microprocessor, one must calculate the total capacitance of an inverter. This capacitance is the sum of the two parallel capacitances, namely the driver and the load capacitance [46].

$$C = C_{load} + C_{driver} \tag{4.22}$$

Each of these components can further be segregated into several components. Figure 4.11 shows different capacitances in a *MOS* transistor structure. Using these capacitances, the driver and load capacitances can be expressed as [46]:

$$C_{driver} = (C_{kja} + C_{kjp} + C_{kjpg}) \cdot W_N + C_{ovw-N} \cdot W_N +$$

$$(C_{kja} + C_{kjp} + C_{kjpg}) \cdot W_P + C_{ovw-P} \cdot W_P \tag{4.23}$$

$$C_{load} = fanout \cdot (3C_{ovw} + C_{ox}) \cdot W_N + fanout \cdot (3C_{ovw} + C_{ox}) \cdot W_P \tag{4.24}$$

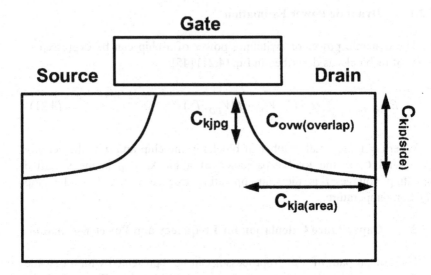

Figure 4-11. Capacitances associated with a *MOS* transistor.

In Eq. (4.23) and Eq. (4.24), different components can be described as follows:

- C_{kja}: Diffusion to substrate capacitance.
- C_{kjp}: Sidewall capacitance.
- C_{kjpg}: Capacitance between Source/Drain diffusions.
- C_{ox}: The gate (oxide) capacitance.
- C_{ovw}: The overlap capacitance between the gate and the Source/Drain.

The factor of 3 before C_{ovw} in Eq. (4.24) accounts for the effective Miller capacitance between the gate and the drain of the *NMOS* and the *PMOS* transistors when the load inverter input makes a positive or negative transition.

For any given block i, the capacitance can be expressed as:

$$C_i = C_{dN} \cdot W_N + C_{dP} \cdot W_P + C_{int} \qquad (4.25)$$

Where W_N and W_P are the total *NMOS* and *PMOS* transistor widths in the block and C_{int}, the contribution of interconnects to the total capacitance in the block is given as:

$$C_{int} = f_1 \cdot (C_{dN} \cdot W_N + C_{dP} \cdot W_P) \qquad (4.26)$$

Where f_I is percentage of interconnect capacitance added to total capacitance in power calculation and:

$$C_d = \beta \cdot (C_{kja} + C_{kjp} + C_{kjpg}) + 2C_{ovw} + C_{ox} \qquad (4.27)$$

Where β is the fitting parameter ($1 \leq \beta \leq 2$), and C_{ox} is the gate capacitance.

7.2.3 Leakage Power

Although leakage power can be attributed to both subthreshold and gate leakage, the primarily focus is on the subthreshold leakage. This is due to the fact that the gate leakage is highly process dependent and can also be tuned to a desirable level by suitable process adjustments. Moreover, for 0.13μm technology the gate leakage power is a small percentage of the subthreshold leakage power (less than 1%) but as the technology scales down, the gate leakage will be a larger portion of the total leakage.

It must be noted that subthreshold leakage is also strongly affected by process variations. Process variations cause variability in the transistor channel length, which in turn causes a variation in the transistor threshold voltage (V_{TH}) mainly due to short-channel effects. The variation in the value of V_{TH} causes a variation in the subthreshold leakage, which can be expressed as [47]:

$$\Delta I_{off} = I_{off-ref} \cdot 10^{\frac{(V_{TH-ref} - V_{TH})}{S}} \qquad (4.28)$$

In Eq. (4.28) V_{TH-ref} and $I_{off-ref}$ are the threshold voltage and leakage current respectively at some reference technology node, and S is the subthreshold swing. This indicates that, for a given temperature, as the threshold voltage decreases, the leakage current increases exponentially.

To include the impact of intra-die threshold voltage or channel length variations, it is necessary to consider the entire range of subthreshold leakage currents, not just the mean or the worst case subthreshold leakage. Let us assume that the intra-die threshold voltage or channel length variations follow a normal distribution for a given transistor width, with μ being the mean and σ being the standard deviation of the distribution. Let I^o be the subthreshold leakage of the device with the mean threshold voltage and mean channel length. Then, by performing the weighted sum of devices of different subthreshold leakage, one can predict the total subthreshold leakage of the chip. This is achieved by integrating the threshold voltage or channel length distribution multiplied by the subthreshold leakage as follows [45]:

$$I_{off} = \frac{I^{\circ}w}{k} \cdot \frac{1}{\sigma\sqrt{2\pi}} \int_{x_{mean}}^{x_{max}} \exp\left(\frac{-(x-\mu)^2}{2\sigma^2}\right) \cdot \exp\left(\frac{(\mu-x)}{a}\right) \cdot dx \quad (4.29)$$

In Eq. (4.29), the first exponent predicts the fraction of the total width for the device subthreshold leakage predicted by the second exponent. If the distribution considered within-die is threshold voltage variation, then x in the above equation represents threshold voltage and a will be equal to $n\Phi_t$. Φ_t is the thermal voltage and n is $1+(C_d/C_{ox})$ [48]. If the distribution considered is channel length, then x in the Eq. (4.28) will represent channel length L, and a will be equal to λ, which can be predicted for a technology by measuring the relationship between the channel length and the device subthreshold leakage. In this study, within-die variation is assumed to be channel length variation. Using Eq. (4.29) and assuming within-die channel length variation the leakage power can be expressed as [45]:

$$\sum_{i=1}^{n} P_i = \sum_{i=1}^{n} \frac{I_{off}^{3\sigma}(V_{DD},T_j)}{m} \cdot \exp\left[\frac{\sigma^2}{2\lambda^2(V_{DD},T_j)} - \frac{3\sigma}{\lambda(V_{DD},T_j)}\right] \cdot W_i \cdot X_n \cdot V_{DD}$$

$$(4.30)$$

Where:
- n is the number of blocks.
- m is fraction of off devices.
- X_n is the noise factor I_{off}.
- σ is the standard deviation of I_{off} process distribution.
- λ is the slope of the I_{off} vs. channel length (L) curve.
- W_i is the total width of transistors in block i.

7.3   Reliability and Cooling Constraints

This section will describe that how the gate oxide reliability and cooling constraints are integrated into the electro-thermal tool.

7.3.1   Reliability Constraints

In the electro-thermal modeling of the chip, it is necessary to consider the long term gate oxide reliability of the chip. In any given junction temperature there is a maximum V_{DD}, beyond which, the gate oxide reliability of the chip will be compromised. To validate the V_{DD} values in the self-consistent methodology, a gate oxide reliability constraint equation is used as given below:

$$V_{DD} \leq V_{max} = T_j \times R + c \tag{4.31}$$

$$R = \frac{xmV}{1^\circ C} \tag{4.32}$$

In Eq. (4.31) and Eq. (4.32), R is a technology dependent reliability factor. The constant c can be calculated based on R and nominal values of V_{DD} and T_j. V_{max} is the maximum voltage for V_{DD} that satisfies the reliability criterion for gate oxide. The reliability criterion is checked at the end of each iterative loop in Figure 4.7 in calculating T_j and V_{DD}.

7.3.2   Cooling Constraints

The operating junction temperature of a chip depends on the cooling solution that is used for conducting the generated heat from the junction to the ambient surrounding the chip. Different cooling solution can be used to remove the generated heat. In this study we focus on the air cooling and refrigeration to compare the low temperature and high temperature operation of microprocessors. The model that is used in the electro-thermal tool is shown in Figure 4.12. In this model:

$$T_j - T_{amb} = P_{chip} \cdot R_{ja} \tag{4.33}$$

$$R_{ja} = R_{jc} + R_{ca} \tag{4.34}$$

$$T_{amb} - T_{out} = P_{sys} \cdot R_{sys} \tag{4.35}$$

$$P_{cooling} = Q_{elec} = \frac{P_{chip}}{COP} \tag{4.36}$$

$$\eta = 1 - \frac{1}{COP} \tag{4.37}$$

$$P_{sys} = P_{chip} + P_{cooling} = P_{chip} + (1-\eta) \cdot P_{chip} = (2-\eta) \cdot P_{chip} \tag{4.38}$$

In Eq. (4.35) and Eq. (4.38), P_{sys} is the total system power which includes total chip power (dynamic and static) as well as power spent due to any dynamic cooling mechanism with an efficiency of η ($\eta < 1$). T_{amb} is the ambient temperature (temperature immediately outside the chip case) and T_{out} is the external room temperature. In these equations, R_{jc} is the junction to case and R_{ca} is the case to ambient thermal resistance. Q_{elec} is the amount of power used by the cooling system and the coefficient of performance (*COP*) is the ratio of the chip power to cooling power. T_a and T_j are ambient and junction temperature respectively.

7.4   Results of Electrothermal Analysis Tool

To demonstrate how the modeling works, Figure 4.13 shows the results of an optimization for an example microprocessor in a low-leakage 130nm process technology for a typical package and air cooling system.

Solutions are obtained for different V_{DD} values, and an operating point is accepted only if the V_{DD} does not exceed the gate oxide reliability-limited maximum supply voltage (V_{max}) at the final T_j . Therefore, reliability considerations set the maximum allowable supply voltage. The highest optimal frequency and corresponding T_j are set by $V_{DD} = V_{max}$ at that T_j. For these simulations, the ambient temperature (T_a) is set to 35°C.

The *X*-axis of Figure 4.13 represents the chip operating frequency. The *Y*-axis has captured multiple parameters including power, temperature, supply voltage and the reliability maximum allowed supply voltage (V_{max}). As the supply voltage increases, the chip frequency increases, and the junction temperature rises. However, the maximum reliability-limited supply voltage reduces at higher temperatures due to degraded gate oxide reliability performance. Consequently the maximum supply voltage is determined at the intersection of the V_{DD} and reliability limited maximum V_{DD} (V_{max}) curves. This sets the junction temperature, frequency and power of the chip accordingly. The optimal operating frequency is 2.7GHz at V_{DD} of 1.5V and T_j of 81°C where the system power is 82W.

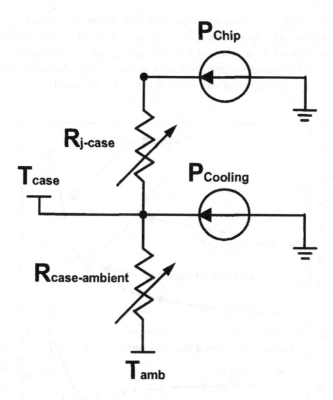

Figure 4-12. Thermal circuit illustrating the relationship between the chip and the system level power, thermal impedance and temperature.

Interconnect RC delays with repeaters can also limit the maximum frequency (top portion of Figure 4.13). RC delays change with T_j in a different way from transistor performance and circuit delay. Also, RC delay is relatively insensitive to V_{DD} change, whereas circuit delays in logic paths change significantly with V_{DD}. In an optimum design interconnect should not limit chip frequency and power at the optimum frequency. This allows transistors to provide their highest potential performance. This is shown in top portion of Figure 4.13 where interconnect (dashed line) increases power without improving the chip frequency after optimal operating point.

8.   SUMMARY

As VLSI circuits are scaled into deep submicron technologies the power consumption of the *ICs* is increased which in turn increases the junction temperature of *ICs*. Prediction of the junction temperature at an early stage of the design is an important factor in the determination of performance and long term reliability of the VLSIs. In this chapter fundamentals of the electrothermal modeling at architecture and circuit were described and a case study with regard to electrothermal modeling was presented.

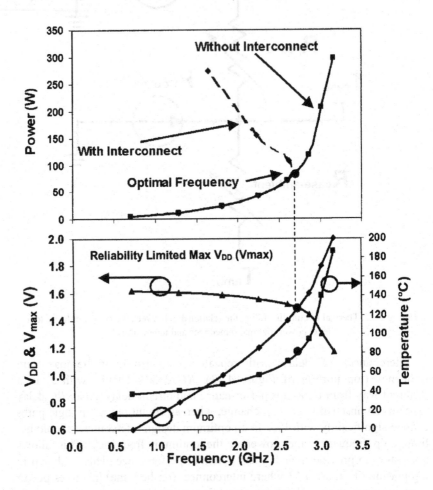

Figure 4-13. Optimization of microprocessor operating frequency subject to reliability constraints.

References

1. Databook "Leistungshalbleiter" (02.97), Siemens AG, S. 115ff.
2. M. Marz, P. Nance, "Thermal Modeling of Power-electronic Systems", Infineon Technologies, AG, Munich.
3. W. Wondrak, "Physical limits and lifetime limitations of semiconductor devices at high temperature". Microelectronics Reliability, Vol. 39 No. 6-7, pages 1113-1120, 1999.
4. International Technology Roadmap for Semiconductors (ITRS). http://public.itrs.net.
5. H. Sanchez, B. Kuttanna, T. Olson, M. Alexander, G. Gerosa, R. Philip, and J. Alvarez. "Thermal management system for high performance PowerPC microprocessors". Proceedings of IEEE COMPCON, pages 325-330, 1997.
6. G. Gerosa, M. Alexander, J. Alvarez, C. Croxton, M. D'Addeo, A.R. Kennedy, C. Nicoletta, J.P. Nissen, R. Philip, P. Reed, H. Sanchez, S.A. Taylor, and B. Burgess, "A 250-MHz 5-W PowerPC microprocessor with on-chip L2 cash controller". IEEE Journal of Solid-State Circuits, Vol. 32, No. 11, pages 1635-1649, 1997.
7. L. Geiling, "Über die elektrische Nachbildung von Wärmeleitungsvorgängen", Siemens Zeitschrift 35, pages 98-104, 1961.
8. D. Brooks and M. Martonosi, "Dynamic thermal management for highperformance microprocessors", In Proceedings of the Seventh International Symposium On High-Performance Computer Architecture, pages 171–82, 2001.
9. W. Huang, J. Renau, S.-M. Yoo, and J. Torellas, "A framework for dynamic energy efficiency and temperature management", In Proceedings Of the 33rd Annual IEEE/ACM International Symposium on Microarchitecture, pages 202–213, 2000.
10. K. Skadron, T. Abdelzaher, and M. R. Stan, "Control-theoretic techniques and thermal-RC modeling for accurate and localized dynamic thermal management", In Proceedings of the Eighth International Symposium on High-Performance Computer Architecture, pages 17–28, 2002.
11. K. Skadrony, M. Stanz, M. Barcellaz, A. Dwarkaz, W. Huangz, Y. Liy, Y. Maz, A. Naiduy, D. Parikhy, P. Rez, G. Rosez, K. Sankaranarayanany, R. Suryanarayanz, S. Velusamyy, H. Zhangz, Y. Zhangz, "HotSpot: Techniques for Modeling Thermal Effects at the Processor-Architecture Level", 8th THERMINIC Workshop, pages 1-4 2002.

12. M. Matson "Circuit implementation of a 600MHZ superscalar RISC microprocessor. Computer Design", VLSI in Computers and Processors, Vol. 26, No. 2, pages 104–110, 1998.
13. S. Lee, S. Song, V. Au, K. Moran, "Constricting/Spreading resistance model for electronics packaging", proceedings of AJTEC, pages 199-206, 1995.
14. K. Skadron, "Temaperature aware micro-architecture: extended discussion and results" Technical report, CS-2003-08, University of Virginia, Department of Computer Science, 2003.
15. Y-K. Cheng, E. Rosenbaum, S-M. Kang, "ETS-A: A New Electrothermal Simulator for CMOS VLSI Circuits", In Proceedings of ED&TC'96, pages 566-570, 1996.
16. Y-K. Cheng, Ch-Ch. Teng, A. Dharchoudhury, E. Rosenbaum, S-M. Kang, "iCET: A complete chip-level thermal reliability diagnosis tool for CMOS VLSI", In Proceedings of ICCAD'96, pages 548-551, 1996.
17. V. Sz'ekely, A. Poppe, A. P'ahi, A. Csendes, and G. Hajas, "Electrothermal and logi-thermal simulation of VLSI designs", IEEE Transaction On VLSI Systems, Vol. 5, No. 3, pages 258–69, 1997.
18. V. Székely, "THERMODEL: a tool for compact dynamic thermal model generation", Proceedings of the 2nd THERMINIC Workshop, pages 21-26, 1996.
19. V. Székely, "Identification of RC Networks by Deconvolution: Chances and Limits", IEEE Transactions on Circuits and Systems-I. Theory and applications, Vol. 45, No. 3, pages 244-258, 1998.
20. V. Székely, "SUNRED: a new thermal simulator and typical applications", Proceedings of the 3rd THERMINIC Workshop, pages 84-90, 1997.
21. M. Rencz, V. Székely, "A generic method for thermal multiport model generation of IC packages", Proceedings of SEMI-THERM pages 145-152, 2001.
22. V. Székely, K. Tarnay, "Accurate algorithm for temperature calculation in nonlinear circuit analysis", Electronics Letters, Vol.8, No. 19, pages 470-472, 1972.
23. K. Nemeth, "On the Analysis of Nonlinear Networks Considering the Effect of Temperature", IEEE Journal of Solid-State Circuits, SC-Vol. 11, No. 8, pages 550-552, 1976.
24. K. Fukahori, P.R. Gray, "Computer simulation of integrated circuits in the presence of electro-thermal interaction", IEEE Journal of Solid-State Circuits, SC- Vol. 11, No. 12, pages 834-846, 1976.
25. V. Székely, "Accurate calculation of device heat dynamics: a special feature of the TRANS-TRAN circuit analysis program", Electronics Letters, Vol. 9, No. 6, pages 132-134, 1973.

26. W.V. Petegem, B. Geeraerts, W. Sansen, B. Graindourze, "Electro-thermal simulation and design of integrated circuits", IEEE Journal of Solid-State Circuits, SSC- Vol. 29, No. 2, pages 143, 1994.

27. W.H. Kao, W.K. Chu, "ATLAS: An Integrated Thermal Layout and Simulation System of ICs", In Proceedings of ED&TC'94, 1994.

28. S. Lee, D.J. Allstot, "Electro-thermal simulation of integrated circuits", IEEE Journal of Solid-State Circuits, SSC- Vol. 28, No. 12, pages 1283-1293, 1993.

29. S. Wunsche, C. Clauss, P. Schwarz, F. Winkler, "Electro-thermal circuit simulation using simulator coupling", IEEE Transactions on Very Large Scale Integration (VLSI) Systems, Volume 5, Issue 3, Pages: 277-282, 1997.

30. W. van Petegem, B. Geeraerts, W. Sansen, B. Graindourze, "Electrothermal simulation and design of integrated circuits", IEEE Journal of Solid-State Circuits, Volume 29, Issue 2, Pages 143-146, 1994.

31. C.C. Lee, A.L. Palisoc, J.M.W. Baynham, "Thermal analysis of solid-state devices using the boundary element method", IEEE Transactions on Electron Devices, Volume 35, Issue 7, Pages 1151-1153, 1988.

32. K. Fukahori, P.R. Gray, "Computer simulation of integrated circuits in the presence of electrothermal interaction", IEEE Journal of Solid-State Circuits, Volume 11, Issue 6, Pages 834-846, 1976.

33. G. Digele, S. Lindenkreuz,; E. Kasper, "Fully coupled dynamic electro-thermal simulation", IEEE Transactions on Very Large Scale Integration (VLSI) Systems, Volume 5, Issue 3, Pages 250-257, 1997.

34. M. Latif, P.R. Bryant, "Network Analysis Approach to Multidimensional Modeling of Transistors Including Thermal Effects", IEEE Transactions on Computer-Aided Design of Integrated Circuits and Systems, Volume 1, Issue 2, Pages 94-101, 1982.

35. A. Haji-Sheikh, "Peak temperature in high-power chips", IEEE Transactions on Electron Devices, Volume 37, Issue 4, Pages 902-907, 1990.

36. X. Gui, G.-B. Gao, H. Morkoc, "Simulation study of peak junction temperature and power limitation of AlGaAs/GaAs HBTs under pulsed and CW operation", IEEE Electron Device Letters, Volume 13, Issue 8, Pages 411 – 413, 1992.

37. K. Chang-Woo, N. Goto, K. Honjo, "Thermal behavior depending on emitter finger and substrate configurations in power heterojunction bipolar transistors", IEEE Transactions on Electron Devices, Volume 45, Issue 6, Pages 1190-1195, 1998.

38. L.L. Liou, B. Bayraktaroglu, "Thermal stability analysis of AlGaAs/GaAs heterojunction bipolar transistors with multiple emitter

fingers", IEEE Transactions on Electron Devices, Volume 41, Issue 5, Pages 629-636, 1994.

39. L. Arthur, A.L. Palisoc, C. Lee Chin," Exact thermal representation of multilayer rectangular structures by infinite plate structures using the method of images", Journal of Applied Physics, Volume 64, Issue 12, pages 6851-6857, 1988.

40. N. Rinaldi, "Thermal analysis of solid-state devices and circuits: an analytical approach", Journal of Solid State Electronics, Volume 44, pages 1789-1798, 2000.

41. DH. Smith, A. Fraser, and. J. ONeil, "Measurement and prediction of operating temperatures", Proceedings of SEMITHERMIN Symposium, 1986.

42. A. Vassighi, A. Keshavarzi, S. Narendra, G. Schrom, Y. Ye, S. Lee, G. Chrysler, M. Sachdev, V. De, "Design optimizations for microprocessors at low temperature", Proceedings of Design Automation Conference, pages 2-5, 2004.

43. K. Banerjee, L. Sheng-Chih, A. Keshavarzi, S. Narendra, and V. De, "A self-consistent junction temperature estimation methodology for nanometer scale ICs with implications for performance and thermal management", IEEE International Electron Devices Meeting, pages 36.7.1-36.7.4, 2003.

44. A. Vassighi, O. Semenov, and M. Sachdev, "Thermal Runaway Avoidance", IEEE International Reliability Physics Symposium, pages 655-656, 2004.

45. S. Narendra, V. De, S. Borkar, D. A. Antoniadis, and A.P. Chandrakasan, "Full chip subthreshold leakage power prediction and reduction techniques for sub-0.18μm CMOS", IEEE Journal of Solid-State Circuits, Vol. 39, No. 3, pages 501-510, 2004.

46. J. M. Rabaey. "Digital Integrated Circuits", Prentice Hall, U.S.A, 1996.

47. S. M. Sze, "Physics of Semiconductor Device", John Wiley & Sons, Inc., U.S.A, 1936.

48. Y. Taur, T.H. Ning, "Fundamentals of modern VLSI devices", Cambridge, UK, Cambridge University Press, 1998.

Chapter 5

THERMAL RUNAWAY AND THERMAL MANAGEMENT

Abstract: This chapter begins with the concept of the thermal runaway and optimization methods to prevent the thermal runaway during reliability screening test (burn-in). Then we review the thermal management of the VLSI circuits during normal operating conditions to prevent thermal runaway and finally discuss the temperature measurement methods.

Key words: Thermal   Runaway,   Thermal   Management,   Junction   Temperature Measurement.

1.   THERMAL AWARENESS

The total power consumption of a high performance microprocessor increases with scaling. The increased power in scaled technologies also increases the temperature of the chip. The safe operation of the chip requires the prevention of excessive chip temperature. Very sophisticated cooling solutions are used to remove the generated heat from the chip. Any small interruption in the operation of the cooling system or any unfavorable change of the ambient temperature may cause overheating and can lead to permanent degradation of the chip or even permanent damage to the chip.

The off-state leakage current under the nominal operating conditions is becoming increasingly large fraction of the total current at 130 nm and sub-100 nm nodes. The off-state leakage is a strong function of the temperature, hence under the burn-in conditions; ratio of the leakage power to the active power becomes adverse. Typically, clock frequencies are kept in tens of *MHz* range during burn-in, which results in substantial reduction in the active power. On the other hand, the voltage and temperature stresses cause the off state leakage power to be the dominant power component. Stressing during burn-in accelerates the defect mechanisms responsible for early life failures. Thermal and voltage stresses increase the junction temperature resulting in accelerated aging. Elevated junction temperature, in turn, causes leakage to further increase. In many situations, this may produce a positive feedback leading to the thermal runaway which results in a permanent damage to the chip. Such situations are more likely to occur as technology is scaled to the nano-meter regime. Figure 5.1 shows a chip that has gone into the thermal runaway [1]. Thermal runaway increases the post burn-in yield loss dramatically therefore must be avoided at all costs.

In order to avoid thermal damages, special techniques should be applied both during the production testing and during the field operation. To apply these techniques, it is crucial to understand, predict or measure the junction temperature under normal and stress conditions. Junction temperature, in turn, is a function of ambient temperature; junction to case, and case to ambient thermal resistances; and power dissipation. Considering these parameters, one can optimize the operating environment or test (burn-in) conditions to minimize the probability of thermal damages and thermal runaway while maintaining the operating performance and effectiveness of the burn-in test.

2.   THERMAL RUNAWAY

Junction temperature of a chip depends on the power consumption of the chip. When the power in the chip increases, the junction temperature increases. The total power of the chip consists of two parts; dynamic power and static power or leakage power. The dynamic power has a weak negative dependency on the junction temperature. The reason is that the I_{on} of the *CMOS* transistor decreases when the junction temperature increases. On the other hand junction temperature and leakage current are strongly correlated and create a positive feedback mechanism between them. Increasing the junction temperature will increase the leakage current and increased leakage current will further increase the junction temperature. Under burn-in or normal operating conditions, designers try to control the junction temperature by removing the heat from the chip. As long as the rate of heat removal is greater or equal to the rate of heat generation, the junction temperature remains constant at the designed operating point. When the rate of heat generation becomes greater than the rate of heat removal, junction temperature starts to increase and the thermal runaway occurs [2].

The junction temperature (T_j) of an *IC* is defined as the average temperature of the silicon substrate. T_j is a crucial parameter in performance determination, reliability-prediction procedures, and burn-in testing. Under nominal conditions, any increase in T_j from the design point degrades the performance. Under burn-in conditions, any deviation in T_j from what is defined for stress conditions; either over stresses the chip or under stresses the chip and consequently causes long term reliability problems.

Figure 5-1. Test socket destroyed by thermal runaway [1].

The junction temperature of a chip can be expresses as:

$$T_j = T_a + P \times R_{ja} \tag{5.1}$$

Where T_a is the ambient or set point temperature, P is the device total power, and R_{ja} is the junction-to-ambient thermal resistance. The power dissipation can be subdivided into dynamic and leakage components, as:

$$P = P_{dynamic} + P_{leakage} \tag{5.2}$$

$$P_{leakage} = V_{DD} \times I_{leakage} \tag{5.3}$$

$$P_{dynamic} = C \times V_{DD}^2 \times f \tag{5.4}$$

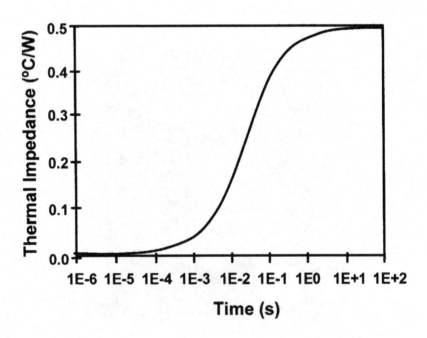

Figure 5-2. Junction to ambient thermal impedance (Z_{ja}) increases with time and reaches its steady-state value of 0.5°C/W [2].

In Eq. (5.4), C is the total *IC* switching capacitance and f is the operating frequency of the chip and can be expressed as:

$$f = \frac{I_{on}}{C \times V_{DD} \times N} \tag{5.5}$$

Where C is the average total capacitance of a node in the critical path and N is the number of logic stages in the critical path.

As it can be seen in Eq. (5.1), thermal resistance of the chip is one of the parameters influencing the junction temperature. Thermal resistance is the resistivity of the material to the heat flow from silicon junction to the ambient surrounding the chip. Thermal resistance depends on the different materials in the chip, package of the chip, heat sink, and cooling system of the chip.

Figure 5.2 shows the transient behavior of the junction to ambient thermal impedance. When the chip is powered on, the thermal impedance starts to increase and reaches to its steady state condition. This phenomenon can also be explained with the help of the thermal equivalent circuit illustrated in Figure 4.1 and Eq. (5.6). At the instance chip is powered on, a large thermal step (large ω) results in smaller thermal impedance. However, in the steady state the value of ω becomes smaller. Therefore the thermal impedance is determined primarily by the thermal resistance under steady state conditions.

$$\frac{1}{Z_{ja}} = \frac{1}{R_{ja}} + j \cdot \omega \cdot C_{ja} \tag{5.6}$$

A typical value for the thermal resistance is 0.5°C/W. Assuming that the generated power must be removed from the chip; the removed power can be derived using Eq. (5.1) as follow:

$$P_{removed} = \frac{T_j - T_a}{R_{ja}} \tag{5.7}$$

It can be seen that in Eq. (5.7) the removed power is a linear function of the junction temperature and the slope of this function is inverse of the thermal resistance of the chip. Figure 5.3 shows the generated power and removed power of a microprocessor as a function of the junction temperature while the ambient temperature is kept at 35°C.

In Figure 5.3, the straight lines represent transient behavior caused by changing thermal impedance with time. As time increases, the thermal impedance increases from 0°C/W and reaches a steady-state value of 0.5°C/W. In other words, the slope of the line which represents the removed power decreases. On the other hand, the exponential curve is the generated power or chip power at the given ambient temperature. An intersection of the straight line (representing the removed power) and the exponential curve (representing the chip power) represents the steady state operating condition of the system where removed heat is equal to the generated heat. As long as there is an intersection between the removed power curve and the chip power curve, thermal runaway will not occur.

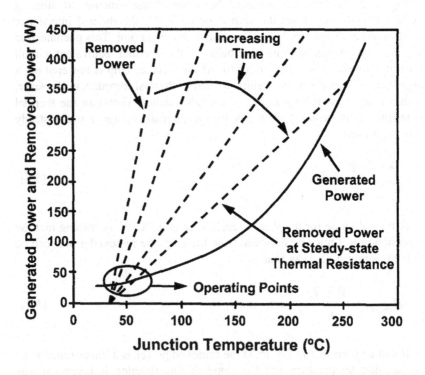

Figure 5-3. 130-nm microprocessor power (exponential), removed power (dashed lines) for a thermal resistance of 0.5°C/W [2].

The concept of the thermal runaway can be further explained with the help of Figure 5.4. The solid curves in the figure represent total generated power in (i) nominal processing conditions, and (ii) when transistors are processed with 10% shorter channel lengths. The dashed lines represent removed power under different steady state thermal resistance values and ambient temperatures. It can be seen that the chip power with nominal channel length has an intersection with the removed power curve at junction temperature of 110°C. The slope of the line is 1/0.5°C/W and the ambient temperature is 80°C. At this operating point the leakage power has fixed value. Now, if owing to increasing activity, the active power, and consequently, the junction temperature is increased beyond 110 °C, since the removed power is larger than the generated power, the operating point converges to the initial operating point. The same is true if the dynamic power is reduced owing to sudden decrease in the activity. The concept of the thermal runaway can be further explained with the help of Figure 5.4. In other words, the junction temperature at the operating point is stable [2].

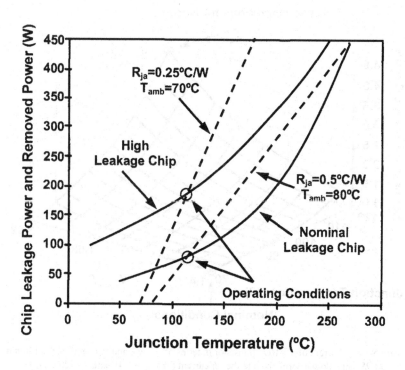

Figure 5-4. Burn-in setup points for nominal leakage and high leakage chips [2].

On the other hand if we look at the curve for the higher power chip, we see that there is no intersection between this curve and the removed power curve with the thermal resistance of 0.5°C/W. Since at all temperatures the removed power is less than the chip power, for this particular chip, the operating conditions will lead the chip to the thermal runaway. To overcome the problem the operating conditions must be changed. The new environment is shown in the figure with a thermal resistance of 0.25°C/W and an ambient temperature of 70°C which creates an intersection between the chip power and the removed power. From this experiment, it can be concluded that for scaled chips with higher power, the setup for the operating conditions must evolve by reducing either the ambient temperature or the thermal resistance or a combination of both of them. This will shift the removed power curve to the left of the corresponding generated power curve for the *IC* at the designed operating conditions.

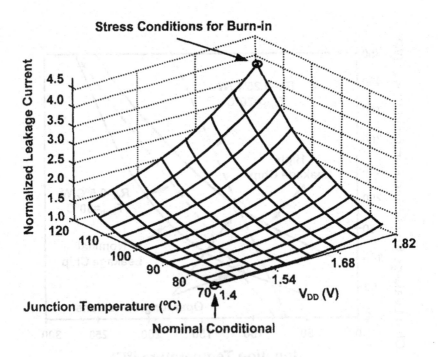

Figure 5-5. Off current of a *NMOS* transistor in terms of voltage and temperature for 130 nm *CMOS* technology (normalized to the *off* current for $V_{DD} = 1.4$V and T_j=120°C) [2].

3. THERMAL RUNAWAY DURING BURN-IN

Burn-in is a crucial reliability screening test to ensure the quality and reliability of products before they are shipped to the customers. Historically, the burn-in environment temperature and voltage have been 125°C and $V_{DD}+30\%$ to $V_{DD}+40\%$, respectively. At the time, the leakage power was a non-issue. However, in sub-180 nm technologies, leakage power is significantly higher under burn-in conditions. Figure 5.5 shows the *NMOS* transistor leakage current increase for 130 nm technology at burn-in conditions. The figure shows the increase in leakage power with increasing temperature and voltage. As it can be seen from the graph, the leakage is increased by approximately 3.5-4x at burn-in conditions compared to nominal conditions.

There are *CAD* tools available to estimate the junction temperature under various environmental conditions [3, 4, 5]. Figure 5.6 depicts the flow chart of one such program. At any initial temperature, the program reads the input current for a single transistor. Based on the circuit implementation and architecture, the total power is computed using Eq. (5.3), Eq. (5.4), and Eq. (5.5) and the junction temperature is updated in Eq. (5.1). Using this procedure [6], for any given voltage and process technology, the junction temperature is calculated and convergence of the obtained temperature is tested [6]. Depending on the result after several iterations the junction temperature will either converge to a temperature or will increase and lead the chip into thermal runaway.

3.1 Simulation Results

Vassighi et al. modeled a 32-bit microprocessor in 130 nm dual-V_{TH} *CMOS* technology to verify the procedure suggested in Figure 5.6 [6]. The parameters of this program were calibrated to the experimental data from an actual microprocessor. These parameters include the burn-in stress voltage and temperature, I_{on} and I_{off} of the *CMOS* transistors, and the layout of the chip.

Figure 5.7 depicts the electro-thermal simulation results for this microprocessor in the burn-in conditions. The solid curves represent the junction temperatures in liquid cooled ovens while the dashed lines represent the junction temperatures in air cooled ovens. As illustrated in Figure 5.7, for air-cooled ovens if the ambient temperature is kept above 10°C the junction temperature starts rising and does not stabilize. This rise in temperature will lead the chip to thermal runaway. The same chip in a liquid cooled burn-in oven (solid lines) will tolerate up to 90°C of ambient temperature without causing thermal runaway. If we want to keep the junction temperature of 110°C then the ambient temperature of liquid cooled oven must be kept at 76°C. The dotted line represents two overlapping curves under air cooled and liquid cooled conditions while maintaining the junction temperature at 110°C (burn in condition).

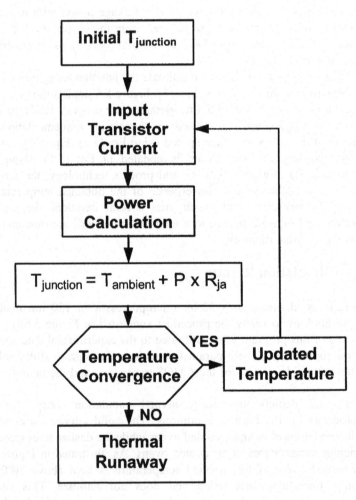

Figure 5-6. A procedure for junction temperature estimation under burn-in conditions [6].

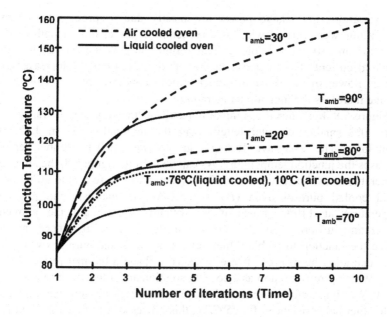

Figure 5-7. Steady-state junction temperature for 130nm high performance microprocessor at burn-in conditions ($T_j = 110°C$ and $V_{DD} = 1.8V$).

Liquid cooled burn-in ovens with a junction to ambient thermal resistance of 0.5°C/W are able to transfer more heat from the chip than air cooled burn-in ovens with junction to ambient thermal resistance of 1.5°C/W. Since the total power at burn-in condition ($T_j=110°C$; $V_{DD}=1.8V$) for this chip is 66W, 1°C/W reduction in thermal resistance will allow us to perform burn-in with 66°C higher ambient temperature. The results in Figure 5.7 confirm that the ambient temperature is increased from 10°C to 76°C in liquid cooled ovens. It should be noted that since the ambient temperature in an air-cooled ovens cannot be less than the a room temperature, it is impractical to burn-in this microprocessor in air-cooled burn-in oven as at room temperature ambient, the chip will eventually go into thermal runaway.

The processors in a production line often have a skewed normal leakage distribution. The processors with larger off state leakage are more susceptible to thermal runaway. Since processors with higher leakage are also faster, the economic cost of losing them to thermal runaway is even higher than the processors with average leakage. Therefore, a flexible burn-in procedure must be tailored according to the leakage. The processors are categorized based on their leakage. Subsequently the burn-in procedure for each category is optimized to minimize the thermal runaway probability.

The variations in leakage power are mostly due to process variations. Simulations were carried with 10% reduction in the channel length of the transistors in 130 nm technology. This resulted in a 3x increase in the sub-threshold current. This increase in the sub-threshold current increases the leakage power of the test chip under burn-in condition by 3x. This extra leakage increases the junction temperature.

Figure 5.8 illustrates the simulation results of the chip that its transistors have 10% smaller channel length than the nominal value in 130 nm technology. The excessive leakage due to the smaller channel length increases the junction temperature. As can be seen in Figure 5.8, the ambient temperature must be reduced from 76°C (shown in Figure 5.7) to 30°C for a liquid cooled burn-in oven ($R_{ja} = 0.5$°C/W) to maintain the junction temperature at 110°C. Since it is difficult to maintain the ambient temperature in burn-in ovens around room temperature, it is necessary to reduce the junction to ambient thermal resistance of the burn-in oven. The next generation burn-in ovens are expected to have a thermal resistance of 0.3°C/W using refrigeration as a cooling solution and a thermal resistance of 0.25°C/W using spray cooling technique as cooling solution, respectively. With a thermal resistance of 0.25°C/W, this processor can be burnt-in in the ambient temperature of 70°C (Figure 5.8).

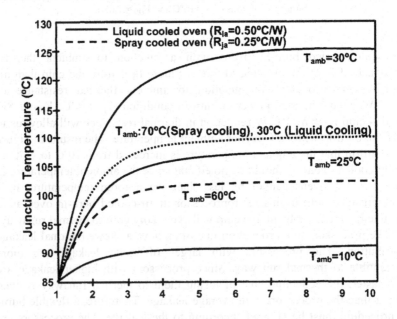

Figure 5-8. Steady-state junction temperature under burn-in conditions for a 130nm high performance microprocessor with a 10% smaller channel length.

4. THERMAL MANAGEMENT DURING NORMAL OPERATING CONDITIONS

The junction temperature during the normal operating conditions depends on the power consumption of the chip, mainly dynamic power consumption. The leakage power during the normal operating conditions is the smaller percentage of the total power. If the junction temperature during the normal operation of the chip is not controlled properly, it will cause long term reliability issues by overstressing of the chip or even in some cases the destruction of the chip. The destruction of the chip is due to thermal runaway. As it was mentioned in section 5.2, the thermal runaway is caused by a positive feedback between the junction temperature and the power consumption of the chip.

Microprocessors due to high power consumption are more susceptible to overstressing and thermal runaway. Microprocessor manufacturers use *Dynamic Thermal Management* (*DTM*) to prevent the thermal overstress and thermal runaway during normal operating conditions. For example, Intel Pentium 4 and Pentium M, the Transmeta Crusoe and Efficeon, the IBM Power5, and the AMD Athlon use some form of *DTM* [7-10]. To explain the dynamic thermal management of high performance microprocessors, we describe the *DTM* of the Intel Pentium 4.

The Pentium 4 microprocessor has two on chip thermal sensors. The chips thermal control unit uses an internal thermal diode which is placed near the part of the *CPU* that is expected to be the hottest part of the chip during the normal operating conditions (hot spot) [7-10]. The internal diode reads the temperature of that part of the chip by converting the amount of the *pn* junction current of the diode to the temperature. When the junction temperature of the microprocessor exceeds a pre-defined temperature, that indicates thermal stress. Under thermal stress the temperature control unit begins throttling *CPU* activity by stopping the *CPU* clock for a period of time. Then if the temperature goes below the pre-defined temperature the unit will start the clock. If the temperature increases, the clock will be stopped again and this process will continue with a controlled duty cycle. The duty cycle can be from 12.5% to 87.5% and is a factory pre-set value (i.e. 2μs for 50% duty cycle).

ACPI (Advanced Configuration and Pow   terface) is an open industry specification co-developed by Hewlett-Packard, Intel, Microsoft, Phoenix, and Toshiba. *ACPI* establishes industry-standard interfaces for OS-directed configuration and power management on laptops, desktops, and servers [11]. An operation system that supports *ACPI* can implement the thermal management policy in three different ways. The first one is incorporating of the active cooling like turning on the fan. The second one is passive cooling where the junction temperature is reduced by reducing the power consumption of the microprocessor. Power is reduced by dynamic voltage scaling (*DVS*) and dynamic frequency scaling (*DFS*) where the V_{DD} and operating frequency (f_{max}) of the microprocessor are scaled according to the junction temperature. The last one is the operation of the shutdown by operating system. The temperature threshold for each policy is set by operating system [12].

The Pentium 4 second on chip sensor unlike the first sensor is a thermal sensor that is software visible. This thermal diode sensor is not located near the thermal control unit sensor and hotspots; therefore the temperature reading of this sensor is not well correlated with the temperature of the hotspots. The sensor diode produces a voltage across two external pins and this voltage is converted to a value using external *A/D* on the motherboard. This value is used by *ACPI* to implement its thermal management policy. It must be noted that this reading represents the nominal junction temperature value and not the junction temperature near the hot spots. Generally, the difference between these two values is known and if one value is measured by sensor, the other one can be computed by the operating system.

5.   TEMPERATURE MANAGEMENT: A CASE STUDY

Today's microprocessors due to their high power consumption and consequently high junction temperature must utilize a thermal management unit to monitor and regulate the temperature of the system [13-16]. If the operating temperature exceeds the upper limit which has been set by the manufacturer, then the thermal management unit forces the system to slow down or even suspend operations temporarily to cool down.

An on-chip thermal management unit is a simple, inexpensive, and low power solution that saves space, power, complexity, and heat-sink costs. The unit provides instantaneous junction temperature readings and as a result enables a thermal management implementation that is tightly coupled to the processor for optimal performance and enhanced reliability. Figure 5.9 shows a block diagram of the on-chip thermal management unit implemented in IBM PowerPC microprocessor [17]. The unit consists of three 32-bit registers, a 2-to-1 multiplexer, a 5-to-32 decoder, an on-chip thermal sensor, an interrupt generator, and a thermal logic control block.

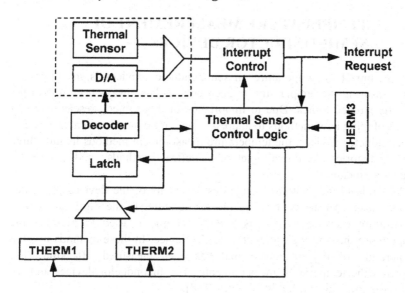

Figure 5-9. The block diagram of the thermal management unit of an IBM PowerPC microprocessor [17].

The thermal management unit provides the following functions:
- Compares the junction temperature with user defined thresholds.
- Generates an interrupt if the temperature exceeds the defined threshold.
- Enables the user to estimate the junction temperature using a software approximation routine.

The unit is accessed and controlled through privileged instructions and unit special purpose registers. These registers are used to configure and control the sensor control logic. THERM1 and THERM2 provide the comparison between the junction temperature and the user specified temperature thresholds. Dual thresholds allow the thermal management software to take different degree of actions when the junction temperature exceeds the threshold. The unit also can operate on a single threshold basis; either THERM1 or THERM2. THERM3 is used to enable the unit and also to control the comparator output sample time. The unit logic manages the interrupt generation in the case of thermal stress as well as other unit control functions [17].

6.   TEMPERATURE MEASUREMENT OF SEMICONDUCTOR DEVICES

The impact of temperature on semiconductor devices is so significant that measuring the temperature of such devices has always been one of the most researched topics. Due to importance of this subject, researchers still are developing new methods to address temperature measurement issues. For semiconductor devices, the temperature measurement methods include three main categories; electrical methods, optical methods, and physically contacting methods.

Methods in the first two categories use some of the devices properties (electrical or optical) to measure the temperature. In the third category, a temperature transducer such as a thermocouple is used to measure the temperature through a close contact to the device. In these methods it is the temperature of the transducer that has been measured. Blackburn has reviewed the temperature measurements of semiconductor devices and we will summarize his work in this section [18].

6.1   Electrical Methods

Some of the electrical parameters of semiconductor devices like *pn* junction forward voltage, threshold voltage, leakage current, and gain are temperature dependent. An accurate measurement of these parameters will be a good indication of the temperature of the device. Although a specific electrical parameter of the device may be related to the temperature of the device at that point, but it will not necessarily represent the temperature of the neighboring areas unless the temperature distribution is uniform in the device. Therefore the distributed nature of the electrical and thermal parameters is lumped into a model with a single temperature, voltage, and current density. Figure 5.10 illustrates the idea of the lumped model [18]. Left hand column in the figure graphically represents temperature and current density of a distributed diode. The right hand column lumps this information into a single diode with constant temperature and constant current density. The disadvantage of this model is that the spatial resolution can not be well determined and only one single, averaged temperature is measured for a device that has some temperature distribution. Despite some inaccuracy of electrical methods, their advantage is that they do not require any special preparation because all the necessary connections are already available.

A major concern with utilizing the electrical parameters for temperature measurement is the separation of the effects of temperature from inherent electrical variation of these parameters. For example, in a diode, when the current is increased, the temperature is increased due to power dissipation in the device. The diode voltage on one hand tends to increased due to increased current. On the other hand the diode voltage tends to decrease due to increased temperature. Therefore the electrical effect must be isolated to utilize the temperature variation of the diode voltage as a thermometer for the diode.

Another concern with electrical measurement is calibration process. To calibrate the electrical parameters that are used as thermometer, the device is set to a series of temperatures using an oven or a hot plate and the electrical parameters of the device are recorded. It is assumed that the temperature of the device is the same as the temperature of the oven or the hot plate. This will cause some inaccuracy due to self heating effect of the device, since there is always some heat dissipation in the device when it is electrically active. In the following sections different electrical measurement methods are reviewed.

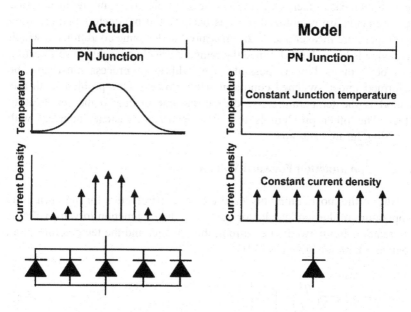

Figure 5-10. Actual (distributed) and lumped models of the *pn* junction temperature, current density and diode models [18].

6.1.1   Switched and Non-Switched Measurements

To address two concerns that were mentioned in previous section, there are two categories of electrical measurements. The first category is switched or pulsed measurements where the effects of the self heating during calibration are minimized by keeping the dissipated power very low. Since the measurement and the calibration must be performed under same electrical conditions, the device dissipated power must be momentarily reduced (switched) to this low value. One problem with switched measurement is that during switching event, transient electrical signals interfere with the measurement. It takes a delay time for the electrical signals to settle down to those of the calibration process. These electrical transients are due to measurement circuit response time, charge storage in the device under test, and parasitic effects of the device package. The other problem with this method is that during the transition of the electrical signals the device temperature decreases due to decreased device power. These two problems cause under estimation of the actual temperature. Methods have been developed to compensate for the effects of electrical transitions to provide better estimation of the temperature [19-21].

The second category of electrical measurements is the non-switched or continuous measurements where the effects of electrical interference during measurements are minimized. This is because the parameters that are used for temperature measurements are measured under same conditions in which the device is being heated. Since the temperature changes due to self heating must be kept as low as possible, the calibration process must now be performed under switched conditions (low power). One problem with this method is the temperature rise during calibration due to the self heating effect. The other problem is that the electrical transients interfere with switching during calibration.

6.1.2   *pn* Junction Forward Voltage

One of the most commonly used electrical parameters for measuring the temperature of the semiconductor devices is the *pn* junction forward voltage. The relationship between the current, the voltage, and the temperature of an ideal *pn* junction is given as [22]:

$$I_{pn} = I_s \left[\exp\left(\frac{qV_{pn}}{kT} \right) - 1 \right]$$

(5.8)

Where I_{pn} is the current of the *pn* junction, q is the electron charge $(1.6x10^{-19}C)$, k is the Boltzmann's constant $(1.381x10^{-23}J/K)$, V_{pn} is the voltage across the junction, and I_s is given as follow:

$$I_s = I_0 T^\gamma \exp\left(\frac{-E_g}{kT}\right) \tag{5.9}$$

Where γ is a constant (around 3), E_g is the bandgap of the *Si* (*1.2eV* at *T=275K*), and I_o is a temperature independent constant. From Eq. (5.8) and Eq. (5.9), the *pn* junction voltage variation at a given current I_{pn} can be derived as follow:

$$\left[\frac{\partial V_{pn}}{\partial T}\right]_{I_{pn}} = -\gamma\frac{k}{q} + \frac{\left(V_{pn} - E_g/q\right)}{T} \tag{5.10}$$

The rate of pn junction voltage change with respect to temperature calculated from Eq. (5.10) is approximately -2mV/K and varies by 7% from 275K to 475K. Some examples of *pn* junction forward voltage as thermometer can be found in [23-38].

6.1.3   Threshold Voltage

The threshold voltage is another temperature dependent electrical parameter. The variation of the threshold voltage with temperature is given by following equation [39]:

$$\frac{dV_{TH}}{dT} = \frac{d\Psi_B}{dT} \cdot \left(2 + \frac{1}{C_{ox}}\sqrt{\frac{\varepsilon_{Si}qN_A}{\Psi_B}}\right) \tag{5.11}$$

Where V_{TH} is the threshold voltage, ψ_B is the distance of the Fermi level from the midgap, ε_{Si} is the dielectric constant of the silicon, N_A is the doping density, E_g is the bandgap at *T=0*, C_{ox} is the gate oxide capacitance and $d\psi/dT$ is given by:

$$\frac{d\Psi_B}{dT} = \frac{1}{T}\left[\frac{E_g(0)}{2q} - |\Psi_B|\right] \tag{5.12}$$

Disadvantage of using threshold voltage as a parameter for temperature measurements is the strong dependency of the threshold voltage on the doping concentration. Therefore any variation on doping concentration will strongly affect the threshold voltage variation and thus it varies from device to device.

6.1.4   Other Electrical Methods

Other methods have been utilized to measure the temperature of the semiconductor devices. Electrical resistance and current gain are among these methods. In electrical resistance method, resistance of a bar is measured by passing a current through it and measuring the applied voltage. Then the changes in measured R can be used as thermometer. In some devices like hetero-structure bipolar transistors (*HBTs*), the gain of the transistor is a monotonic function of the temperature and thus the gain can be used to measure the temperature. However in most devices, the gain is a complex function of the temperature and makes it too complicated to perform temperature measurement using this electrical parameter. Other methods have been developed to measure the temperature of the semiconductor devices which can be found in [40-50].

6.1.5   Functional Description of On-chip Temperature Sensor of PowerPC

The on-chip temperature sensor circuit in IBM PowerPCs is based on the differential voltage change across two diodes biased at the same operating current. Figure 5.11 illustrates the logic flow of the thermal sensor circuitry [17]. One of the diodes is larger than the other one. The voltage across these diodes is sampled at a given temperature. Then the current generated in diodes are compared. When temperature increases, the voltage across the larger diode decreases much more rapidly than the voltage across the smaller diode. Therefore the generated current in larger diode is smaller than the smaller diode. The current from the larger diode is added to reference current which is set based on the thermal threshold of the device. The result is compared to the current from smaller diode to see if the thermal threshold is exceeded.

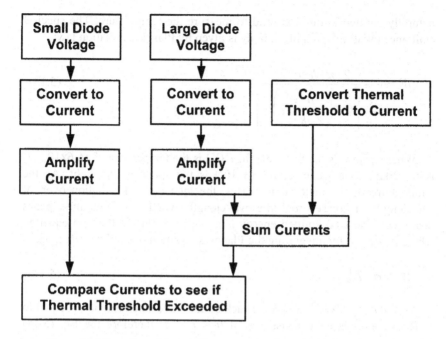

Figure 5-11. Procedure of the on-chip temperature sensor for IBM PowerPC [17].

6.2   Optical Methods

To utilize the optical properties as a thermometer, naturally emitted radiation, reflected radiation, or stimulated emitted radiation is measured. In optical methods, an optical beam of photons is focused at a point on the device and incident photons are reflected after interacting with the device. The properties of the reflected photons are strong function of the local temperature. The changes in reflected beam can be used to obtain the temperature of different parts of the chip. Another commonly used optical technique is the measurement of the naturally emitted infrared radiation from a heated body.

Optical methods have some drawbacks. The device must be accessible to perform the thermal measurements; therefore optical methods are ill suited for such measurements. Moreover, optical methods require expensive, and hard to use equipments [18].

6.2.1   Infrared Imaging

The most commonly used optical method to measure the temperature is the infrared radiation [51-54]. In this method a portion of the spectrum of the naturally emitted infrared is used to measure the temperature. The spectral emittance (Watt/m^3) of a black body is given by Planck's radiation law:

$$W = \frac{2 \cdot \pi \cdot h \cdot c^2}{\lambda^5 \cdot \left[\exp\left(\dfrac{h \cdot c}{\lambda \cdot k \cdot T} \right)^{-1} \right]} \qquad (5.13)$$

Where λ (μm) is the wavelength, T (K) is the temperature, h (6.6×10^{-34} J) is the Planck's constant and c (3×10^8) is the speed of the light. As the temperature of the material increases, the total emitted radiation over all wavelengths increases and the wavelength which the emittance peaks decreases. The total emitted energy is given by Stefan-Boltzmann equation which is derived by integrating the Planck's equation over all wavelengths.

$$W = \sigma \cdot T^4 \qquad (5.14)$$

Where $\sigma = 5.7\times10^{-8}\ Wm^{-2}K^{-4}$. Black bodies have an emissivity, ε, equal to 1. Real materials have emissivity of $0 < \varepsilon < 1$, therefore the Eq. (5.14) becomes:

$$W = \varepsilon \cdot \sigma \cdot T^4 \qquad (5.15)$$

Using Eq. (5.15), by measuring the total emitted energy, the temperature of the material can be determined. The major concern with infrared measurement is the determination of relative permittivity (ε_r). The value of the ε_r for different materials on the chip varies from 0.1 (Aluminum) to 0.6 (polysilicon). Without taking this variation into account, the temperature measurement using infrared method will be impossible. The spatial resolution of infrared method is limited to the wavelength of the detected radiation and is from 3μm to 5μm.

6.2.2   Luminescence

Luminescence is the emission of the radiation due to an external stimulation such as an electric field or a photo excitation. The emitted radiation is the result of electron hole recombination. The source of the electrons and holes is either electrical or optical. In Electroluminescence electrons and holes are injected across a *pn* junction [55]. In Photoluminescence the source of electrons and holes is from external optical excitation [56-58]. Electrons and holes recombine with the peak energy occurring in direct bandgap materials at the bandgap energy. Since the bandgap energy depends on the temperature of the material, the peak energy of the luminescence depends on the temperature. The resolution of the electroluminescence depends on the area of the *pn* junction where the resolution of the photoluminescence depends on the size of the incident optical excitation and can be 0.5μm-1μm and the temperature resolution is 1°C. Luminescence is only useful for direct bandgap semiconductors such as *GaAs*.

6.2.3   Raman

The temperature of *Si MOSFETs*, *GaAs*, and other semiconductor devices has been measured using Raman effect. Raman scattering occurs when a photon scatters from a crystal, with the creation or annihilation of one or more phonons. Since the phonon spectra is temperature dependent, the spectra of the scattered photon is temperature dependent and will be different than the spectra of the incident photon. The spatial resolution of the Raman effect for measuring the temperature is 1μm and the temperature resolution is 1°C -2°C [59, 60].

6.2.4   Reflectance

If the incident and the reflected photons have the same wave length, then the intensity of the reflected photons is temperature dependent. The relative change in optical reflectance density is small and it is in the order of 10^{-5} to 10^{-4} °K^{-1}. Reflectance method as a thermometer has been used to measure the temperature of interconnects, devices, and thin films. For the temperature measurement of self-heated devices and interconnects, a single probe is used. For the temperature measurement of thin films two probes are used; one for heating the thin film and the other for measuring the temperature. Thermo-reflectance is used to measure very short thermal transients (10ns) with the spatial resolution of 1μm [61-63].

6.2.5   Thermo Optic Effect

Thermo optic effects are changes in the optical index of the reflection of a material with the temperature. These changes are small and in the order of 10^{-5} °K^{-1} [64]. This method has been used to measure the temperature profile of the power gate-turn-off thyristor and the temperature in waveguide modulators.

6.3   Physically Contacting Methods

Contacting methods measure the temperature by making thermal contact with the device under study. Contacting methods include single point contact methods, multi-point or blanket cover methods. The spatial resolution of the contacting methods is the smallest of all available methods and is around 30nm to 50nm for scanning thermal probes.

6.3.1   Scanning Thermal Probes

Scanning thermal probes are among the single point contact methods. In this method an atomic force microscope has been modified with a temperature sensitive element such as a thermocouple or thermistor. This method offers the highest spatial resolution due to small size of the probe tip [65].

6.3.2   Liquid Crystals

Liquid crystals are an example of multi point or blanket cover methods. Liquid crystals as an organic materials exist as a phase between a solid and an isotropic liquid. They scatter the incident light based on the wavelength of the light and temperature of the liquid crystal. The liquid crystal molecules have a helical structure and the distance of the molecules is in the order of the wavelength of the visible light. The wave length of the light that is reflected is determined by the pitch of the helix and since the pitch of the helix is temperature dependent, the color of the region being observed changes with temperature of the region. The color range of the liquid crystals can be changed by changing the chemical composition used in the liquid crystal. Liquid crystals are commercially available in the temperature range of -30°C to 120°C, spatial resolution of 1μm, and temperature resolution of 0.1°C [66, 67].

7.   SUMMARY

Thermal management is the key to ensure long term reliability, performance, and thermal runaway avoidance. Unabated increase in power consumption, and poor thermal management during burn-in or even normal operational conditions, may result in thermally induced total chip destruction which is also known as thermal runaway. A constant monitoring and measurement of the junction temperature is significantly important for thermal runaway avoidance. Moreover, this knowledge leads to strategies for optimization of operating conditions.

References

1. Mark Miller (AMD). "Next generation burn-in and test systems for Athlon microprocessors: hybrid burn-in", Burn-in and Test Socket Workshop, 2001.
2. A. Vassighi, O. Semenov, and M. Sachdev. "Thermal Runaway Avoidance", IEEE International Reliability Physics Symposium, pages 655-656, 2004.
3. D. Brooks and M. Martonosi, "Dynamic Thermal Management for High-Performance Micro-processors", Proceedings of 7th International Symposium High-Performance Computer Architecture, IEEE CS Press, pages 171-182, 2001.
4. K. Kanda, K. Nose, H. Kawaguchi, and T. Sakurai, "Design Impact of Positive Temperature dependence on Drain Current in Sub-1-V CMOS VLSIs", IEEE Journal of Solid-State Circuits, Vol. 36, No. 10, 2001..
5. K. Banerjee, L. Sheng-Chih, A. Keshavarzi, S. Narendra, and V. De, "A self-consistentjunction temperature estimation methodology for nanometer scale ICs with implications for performance and thermal management", IEEE International Electron Devices Meeting, pages 36.7.1-36.7.4, 2003.
6. A. Vassighi, O. Semenov, M. Sachdev, and A. Keshavarzi, "Thermal management of high performance microprocessors in burn-in environment", Proceedings of 18th IEEE International Symposium on Defect and Fault tolerance in VLSI Systems, 2003.
7. Intel Corporation, "Intel Pentium 4 Processor with 512-KB L2 Cache on 0.13 Micron Process and Intel Pentium 4 Processor Extreme Edition Supporting Hyper-Threading Technology: Datasheet", Order number 298643-012, 2004.
8. Intel Corporation, "Mobile Intel Pentium 4 Processor-M: Datasheet", Order Number 250686-007, 2003.
9. Intel Corp., "Intel Pentium 4 Processor with 512-KB L2 Cache on 0.13 Micron Process Thermal Design Guidelines: Datasheet", Order Number 252161-001, 2002.

10. Intel Corp., "IA-32 Intel Architecture Software Developer's Manual. Volume 3: System Programming Guide", Order number 253668, 2004.
11. http://www.acpi.info.
12. ACPI4Linux Documentation, http://acpi.sourceforge.net/documentation/thermal.html.
13. Advanced Power Management (APM) Specification by Intel and Microsoft.
14. B. Travis, "Temperature management ICs combat system meltdown", EDN Magazine, pages 39-48, 1996.
15. A. Bakker, J.H. Huijsing, "Micropower CMOS temperature sensor with digital output", IEEE Journal of Solid State Circuits, Vol. 31, No.7, pages 933-937, 1996.
16. K.S. Szajda, C.G. Sodini, H.F. Bowman, "Micropower CMOS temperature sensor with digital output", IEEE Journal of Solid State Circuits, Vol. 31, No.9, pages 1308-1313, 1996.
17. H. Sanchez, B. Kuttanna, T. Olson, M. Alexander, Motorola, G. Gerosa, Motorola, R. Philip, Motorola, J. Alvarez, Motorola, "Thermal Management System for High Performance PowerPCTM Microprocessors", Proceedings of COMPCON pages 325, 1997.
18. D.L. Blackburn, "Temperature measurements of semiconductor devices - a review", Twentieth Annual IEEE Semiconductor Thermal Measurement and Management Symposium, Pages 70-80, 2004.
19. D.L. Blackburn, "A review of thermal characterization of power transistors", Fourth Annual IEEE Semiconductor Thermal and Temperature Measurement Symposium, SEMI-THERM IV., Pages:1-7, 1988.
20. D.W. Berning, and D.L. Blackburn, "The effect of magnetic package leads on the measurement of thermal resistance of semiconductor devices", IEEE transaction on Electron Devices, Vol. ED-28, No. 5, pages 609-611,1981.
21. D.L. Blackburn, F.F. Oettinger, and S. Rubin, "Transient thermal response of power transistors", IEEE Transaction on Industrial Electronics and Control Instrumentation, Vol. IECE-22, No.2, pages 134-141, 1975.
22. S.M. Sze, "Physics of semiconductor devices", John Wiley and Sons, 1981.
23. D.L. Blackburn, "Semiconductor device temperature measurement", Future Circuit International, Vol.4, pages 75-83, 1998.
24. M.G. Adlerstein, M.P. Zaitlin, "Thermal resistance measurements for AlGaAs/GaAs heterojunction bipolar transistors", IEEE Transactions on Electron Devices, Vol.38, Issue 6, Pages 1553 – 1554, 1991.
25. Y.H. Chang, Y.T. Wu, "Measurement of junction temperature in hetrojunction bipolar transistors", Proceedings of 3rd IEEE

InternationalCaracus Conference on Device, Circuits, and Systems, D59/1-D59/4, 2000.
26. N. Bovolon, P. Baureis, J.E. Muller, P. Zwicknagel, R. Schultheis, and E. Zanoni, "A Simple Method for the Thermal Resistance Measurement of AlGaAs/GaAs Heterojunction Bipolar Transistors", IEEE Transactions on Electron Devices, Vol. 45, No. 8, pages 1846-1848, 1998.
27. D.E. Dawson, "CW Measurement of HBT Thermal Resistance", IEEE Transactions on Electron Devices, Vol. 39, No. 10, pages 2235-2239, 1992.
28. W. Liu, "Measurement of Junction Temperature of an AlGaAs/GaAs Heterojunction Bipolar Transistor Operating at Large Power Densities", IEEE Transactions on Electron Devices, Vol. 42, No. 2, pages 358-360, 1995.
29. G.B. Gao, M.S. Unlu, H. Morkoc, and D.L. Blackburn, "Emitter Ballasting Resistor Design for, and Current Handling Capability of AlGaAs/GaAs Power Heterojunction Bipolar Transistors", IEEE Transactions on Electron Devices, Vol. 38, No. 2, pages 185-196, 1991.
30. Guidelines for the Measurement of Thermal Resistance of GaAs FETs, JEDEC Publication No.10, Electronic Industries Association, Washington DC, 1988.
31. R.J. Donarski, "Pulsed I-V and Temperature Measurement System for Characterisation of Microwave FETs", IEEE International Microwave Symposium Digest, pages 1523-1526, 1995.
32. H. Fukui, "Thermal Resistance of GaAs Field-Effect Transistors", IEEE Transactions on Electron Devices, Vol. ED-5, No. 1, pages 118-121, 1980.
33. M. Nishiguchi, M. Fujihara, A. Miki, and H. Nishizawa, "Precision Comparison of Surface Temperature Measurement Techniques for GaAs ICs", IEEE Transactions on Components, Hybrids, and Manufacturing Technology, Vol. 16, No. 5, pages 543-549, 1993.
34. S. Feng, X. Xie, C. Lu, G. Shen, G. Gao, and X. Zhang, "The Thermal Characterization of Packaged Semiconductor Device", Proceedings 16th Annual IEEE Semiconductor Temperature Measurement and Management Symposium, pages 220-226, 2000.
35. Z. Jakopovic, Z. Bencic, and F. Kolonic, "Important Properties of Transient Thermal Impedance for MOSGated Power Semiconductors", Proceedings of the IEEE International Symposium on Industrial Electronics, Vol. 2, pages 574-578, 1999.
36. D.L. Blackburn, and D.W. Berning, "Power MOSFET Temperature Measurements", Proceedings of the IEEE Power Electronics Specialists Conference, pages 400-407, 1982.

37. A. Piccirillo, "Complete Characterisation of Laser Diode Thermal Circuit by Voltage Transient Measurements", Electronics Letters, Vol. 29, No. 3, pages 318-320, 1993.

38. S. Feng, X. Xie, W. Liu, C. Lu, Y. He, and G. Shen, "The Analysis of Thermal Characteristics of the Laser Diode by Transient Thermal Response Method", Proceedings of 5th International Conference on Solid-State and Integrated Circuit Technology, pages 649-652, 1998.

39. B.J. Baliga, "Modern Power Devices", John Wiley and Sons, Inc., 1987.

40. S.C. Cripps, "A New Technique for Screening and Measuring Channel Temperature in RF and Microwave Hybrid Circuits", Proceedings of 6th Annual IEEE Semiconductor Thermal and Temperature Measurement Symposium, pages 40-42, 1990.

41. S.P. Marsh, "Direct Extraction Technique to Derive the Junction Temperature of HBT's Under High Self-Heating Bias Conditions", IEEE Transactions on Electron Devices, Vol. 47, No. 2, pages 288-291, 2000.

42. P.M. McIntosh, and C.M. Snowden, "Measurement of Heterojunction Bipolar Transistor Thermal Resistance Based on a Pulsed I-V System", Electronics Letters, Vol. 33, No. 1, pages 100-101, 1997.

43. J.R. Waldrop, K.C. Wang, and P.M. Asbeck, "Determination of Junction Temperature in AlGaAs/GaAs Heterojunction Bipolar Transistors by Electrical Measurement", IEEE Transactions on Electron Devices, Vol. 39, No. 5, pages 1248-1250, 1992.

44. D.T. Zweidinger, R.M. Fox, J.S. Brodsky, T. Jung, and S.-G. Lee, "Thermal Impedance Extraction for Bipolar Transistors", IEEE Transactions on Electron Devices, Vol. 43, No. 2, pages 342-346, 1996.

45. A.R. Reid, T.C. Kleckner, M.K. Jackson, D. Marchesan, S.J. Kovacic, and J.R. Long, "Thermal Resistance in Trench-Isolated Si/SiGe Heterojunction Bipolar Transistors", IEEE Transactions on Electron Devices, Vol. 48, No. 7, pages 1477-1479, 2001.

46. A. Ammous, "Transient Temperature Measurements and Modeling of IGBT's Under Short Circuit", IEEE Transactions on Power Electronics, Vol. 13, No. 1, pages 12-25, 1998.

47. B.M. Tenbroek, M.S.L. Lee, W. Redman-White, R.J.T. Bunyan, and M.J. Uren, "Self-Heating Effects in SOI MOSFETs and Their Measurement by Small Signal Conductance Techniques", IEEE Transactions on Electron Devices, Vol. 43, No. 12, pages 2240-2248, 1996.

48. W. Redman-White, M.S.L. Lee, B.M. Tenbroek, M.J. Uren, and R.J.T. Bunyan, "Direct Extraction of MOSFET Dynamic Thermal Characteristics From Standard Transistor Structures Using Small Signal Measurements", Electronics Letters, Vol. 29, No. 13, pages 1180-1181, 1993.

49. J. Wei, S.K.H. Fung, W. Liu, P.C.H. Chan, and C. Hu, "Self-Heating Characterization for SOI MOSFET Based on AC Output Conductance",

Proceedings IEEE International Electron Devices Meeting, pages 175-178, 1999.

50. Wei Jin, W.Liu, S.K.H. Fung, P.C.H. Chan, and C. Hu, "SOI Thermal Impedance Extraction Methodology and Its Significance for Circuit Simulation", Transactions on Electron Devices, Vol. 48, No. 4, pages 730-736, 2001.

51. J.P. David, J. Duveau, J. Guerin, and A. Michel, "Electrical and Thermal Testing and Modeling of Breakdown in Space Solar Cells and Generators", Conference Record 23rd IEEE Photovoltaic Specialists Conference, pages 1415-1420, 1993.

52. A. Hefner, D.W. Berning, D.L. Blackburn, and C. Chapuy, "A High-Speed Thermal Imaging System for Semiconductor Device Analysis", Proceedings 17th Annual IEEE Semiconductor Thermal Measurement and Management Symposium, pages 43-49, 2001.

53. J. McDonald, and G. Albright, "Microthermal Imaging in the Infrared", Electronics Cooling, 1997.

54. A. Yasuda, H. Yamaguchi, Y. Tanabe, N. Owada, and S. Hirasawa, "Direct Measurement of Localized Joule Heating in Silicon Devices by Means of Newly Developed High Resolution IR Microscopy", Proceedings of 29th IEEE Annual Reliability Physics Symposium, pages 245-249, 1991.

55. F. Schuermeyer, R. Fitch, R. Dettmer, J. Gillespie, C. Bozada, K. Nakano, J. Sewel, J. Ebel, T. Jenkins, and L.L. Liou, "Thermal Studies on Heterostructure Bipolar Transistors Using Electroluminescence", Proceedings IEEE Cornel Conference on High Performance Devices, pages 45-50, 2000.

56. J.P. Landesman, D. Floriot, E. Martin, R. Bisaro, S.L.Delage, and P. Braun, "Temperature Distributions in III-V Microwave Power Transistors Using Spatially Resolved Photoluminescence Mapping", Proceedings of the 3rd IEEE Caracas Conferences on Devices, Circuits and Systems, D1114/1-D1114/8, 2000.

57. D.C. Hall, L. Goldberg, and D. Mehuys, "Technique for Lateral Temperature Profiling in Optoelctronic Devices Using a Photoluminescence Microprobe", Applied Physics Letters, Vol. 61, No. 4, pages 384-386, 1992.

58. Q. Kim, B. Stark, and S. Kayali, "A Novel, High Resolution, Non-Contact Channel Temperature Measurement Technique", Proceedings of 36th Annual IEEE Reliability Physics Symposium, pages 108-112, 1998.

59. M. Kuball, J.M. Hayes, M.J. Uren, I. Martin, J.C.H. Birbeck, R.S. Balmer, and B.T. Hughes, "Measurement of Temperature in Active High-Power AlGaN/GaN HFETs Using Raman Spectroscopy", IEEE Electron Device Letters, Vol. 23, No. 1, pages 7-9, 2002.

60. J. He, V. Mehrotra, and M.C. Shaw, "Ultra-High Resolution Temperature Measurement and Thermal Management of RF Power Devices Using Heat Pipes", Proceedings of 11th Annual Symposium on Power Semiconductor Devices and ICs, pages 145-148, 1999.
61. Y.S. Ju, and K.E. Goodson, "Thermal Mapping of Interconnects Subjected to Brief Electrical Stresses", IEEE Electron Device Letters, Vol. 18, No. 11, pages 512-514, 1997.
62. R. Abid, and F.-Z. Mezroua, "New Technique of Temperature Noncontact Measurements: Application to Thermal Characterization of GTO Thyristors in Commutation", Canadian Conference on Electrical and Computer Engineering, Vol. 1, pages 586-589, 1995.
63. Y.S. Ju, and K.E. Goodson, "Short-Timescale Thermal Mapping of Interconnects", Proceedings of 35th Annual Reliability Physics Symposium, pages 320-324, 1997.
64. C.C. Lee, T.J. Su, and M. Chao, "Transient Thermal Measurements Using Thermooptic and Thermoelectric Effects", Proceedings of 8th Annual IEEE Semiconductor Thermal Measurement and Management Symposium, pages 41-46, 1992.
65. C.C. Williams, and D. Williams, "Scanning Thermal Profiler", Applied Physics Letters, Vol. 49, No. 23, pages 1587-1589, 1986.
66. P. Jeong, W.S. Moo, and C.C. Lee, "Thermal Modeling and Measurement of GaN-Based HFET Devices", IEEE Electron Device Letters, Vol. 24, No. 7, pages 424-426, 2003.
67. A.M. Chaudhari, T.M. Woudenberg, M. Albin, and K.E. Goodson, "Transient Liquid Crystal Thermometry of Microfabricated PCR Vessel Arrays", Journal of Microelectromechanical Systems, Vol. 7, No. 4, pages 345-355, 1998.

Chapter 6

LOW TEMPERATURE CMOS OPERATION

Abstract: Low temperature operation with utilization of refrigeration as cooling electronic devices. In this chapter we review the low operation of the system has shown improvement in performance and the reliability of *CMOS* devices and describe different cooling solution for low temperature operation.

Key words: Low Temperature, Cryogenic Temperature, Low Temperature *CMOS* Characteristics, Refrigeration.

1. LOW TEMPERATURE MOTIVATION

In the last decade there have been an increasing number of manufacturers offering low temperature cooling of *CMOS* microprocessors to enhance the performance. Although the primary reason for low temperature operation of microprocessors has been performance enhancement, due to increased power in scaled technologies, the low temperature operation may be required to keep the temperature of the microprocessors within accepted limits. Figure 6.1 shows several *CMOS* performance enhancement techniques including low temperature operation [1, 2].

Virtually all commercial computers were designed to operate at temperatures above ambient. The term low temperature refers to any junction temperatures lower than temperatures of air-cooled chips, which typically operate in the range of 60°C to 100°C. A number of cooling technologies are available for providing low temperature cooling with a wide variety of available options. These cooling options and their temperature ranges are shown in Figure 6.2 [1].

With the increase in *CMOS* performance achieved with lower temperature, a number of companies have started programs to investigate cooling electronics. DEC, AMD, IBM, Sun Microsystems, SYS technologies, and Kryotech Inc are among companies that have developed computers with utilization of vapor compression refrigeration cycle [3]. Since refrigerators are widely available, and the compressor and the fan are the only moving parts in the cooling system, the building blocks embody a stable, reliable and mature technology.

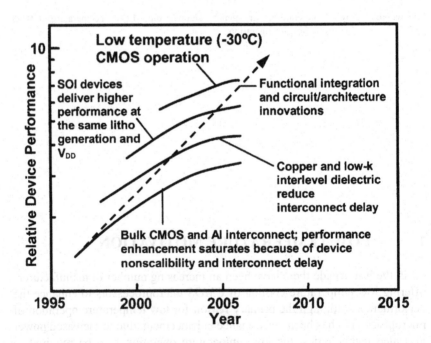

Figure 6-1. Application of new structures, materials, and techniques to maintain continuation of Moore's law (dashed curve) [1].

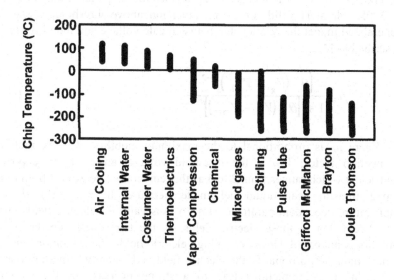

Figure 6-2. Electronic cooling techniques and their corresponding operating temperature ranges.

2. LOW TEMPERATURE CHARACTERIZATION OF CMOS DEVICES

Lowering the operating temperature introduces some benefits in scaled CMOS technologies. Among them are substantial increase of the mobility and the saturation velocity, better turn-on capabilities, latch-up immunity, lower chip power consumption, and reduced thermal noise. However, there are some drawbacks related to low temperature operation especially some cryogenic conditions. In this section some of the *CMOS* device characteristics will be reviewed.

2.1 Mobility

The mobility of carriers in surface channel *MOS* transistors is one of the important parameters that are needed to characterize the behavior of the device. The mobility law is of the prime importance to describe the transfer and output characteristics of a *MOS* device. As the temperature is decreased, the mobility model is modified to include the effects of low temperature. A generalized mobility model has been proposed for temperature range of 4.2K to 300K which is valid for strong inversion above threshold. Using the generalized model the relationship between gate voltage and mobility can be described as [4, 5]:

$$\mu = \mu_m \frac{\left[\theta \cdot \left(V_g - V_{TH}\right)\right]^{n-2}}{1 + \left[\theta \cdot \left(V_g - V_{TH}\right)\right]^{n-1}} \qquad (6.1)$$

Where μ_m is proportional to the maximum mobility, θ is the mobility attenuation factor, V_{TH} is the threshold voltage, and n is an exponent coefficient which ranges from 2 to 3 as the temperature decreased from room temperature to near liquid helium temperature. In Eq. (6.1), θ is a temperature dependent parameter. Eq. (6.1) is valid for electric fields up to 3-4MV/cm. Up to these electric fields, as the temperature decreases the mobility is increased. However, in high electric fields, the saturation velocity is no longer proportional to the electric field and velocity saturation due to optical phonon interactions lead to a strong decrease of the mobility irrespective of the temperature [6, 7]. Figure 6.3 shows the trans-conductance of the *N*-channel MOS transistor with gate voltage for various temperatures.

In low electric fields, the mobility increases by the factor of 4-6 when the temperature is decreased from 300°K to 77°K because of the reduced carrier scattering due to lattice vibrations. Figure 6.4 shows the temperature dependence of the channel mobility for both electrons (*NMOS*) and holes (*PMOS*), measured in the linear mode on a 0.5μm *CMOS* technology. This increase in mobility is also observed on long channels in the saturation mode, where it can be assumed that drift velocity in the channel is proportional to longitudinal electric field [8].

Another important point that should be mentioned is the difference between mobility of the electrons and holes near the threshold voltage. As the temperature decreases, the mobility of the electrons is increased monotonically. However, the mobility of the holes has a different behavior depending on whether the channel is in strong or weak inversion. In strong inversion, the mobility increases with the temperature decrease. On the other hand, in weak inversion initially the mobility increases with temperature decrease and then at very low temperature the mobility decreases with the decrease in temperature [9].

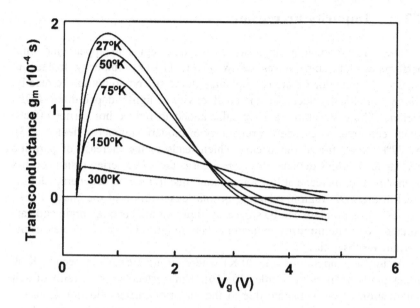

Figure 6-3. The transconductance variation of the *N*-channel transistor with gate voltage for different temperatures.

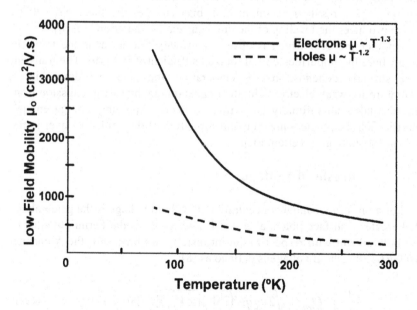

Figure 6-4. Increase in mobility in low fields when the temperature decreases [8].

2.2   Impurity Freeze out

The effect known as impurity freeze out begins to appear and limit performance at temperatures below 100°K [10-12]. Impurity freeze out occurs when the dopant atoms implanted in the silicon do not ionize readily, which dramatically decreases the conductivity of lightly doped drain (*LDD*) regions. These structures are typically added to reduce hot carrier effects, which can greatly degrade the reliability of modern submicron devices [13]. At 77°K weak freeze out occurs which leads to the increase of parasitic resistance in *MOS* devices. In such a case, the *LDD* series resistance has been shown to increase at low temperature due to the non degenerate doping level in the *LDD* region, resulting in total deactivation of the channel [14]. However this problem can be solved by applying high enough drain and gate voltages to create impurity ionization. The impurity ionization decreases the resistance substantially.

At lower temperatures (T<30°K) where strong freeze out occurs, Kink effect plays an important role [15, 16]. Kink effect is as a result of self polarization of the substrate due to the majority carriers flowing from the body to the source. The impurity freeze out causes a strong increase of the back resistance and prevents the evacuation of the drain impact ionization current through the body contact. This gives rise to the self biasing of the body and therefore increases the forward polarization of the source-substrate junction. As a result the shift in bulk bias changes the threshold voltage which produces the leveling of the drain current in saturation [15-18].

Another problem which is observed in deep freeze out is the transient time where the impurities do not ionized sufficiently fast [19]. The transient time strongly depends on the temperature and gate and drain biases. Therefore for weak electric field, the ionization rate by carrier emission from the trap takes an infinitely large time to occur. Applying a large enough electric field helps the ionization and formation of the depletion layer in the silicon substrate after certain time.

2.3   Threshold Voltage

For a uniform channel concentration, threshold voltage is the gate voltage that creates a surface potential $\psi_s \approx 2\Phi_B$, where Φ_B is the Fermi potential of the bulk silicon with respect to the intrinsic Fermi level. For the *N*-channel *MOSFET*, threshold voltage is defined as [8]:

$$V_{TH} = \frac{1}{C_{ox}}\left[-Q_{eff} + \sqrt{2\varepsilon_{Si}qN\left(2\left|\Phi_B\right| \pm V_{SB}\right)}\right] + 2\left|\Phi_B\right| + \Phi_{ms} \qquad (6.2)$$

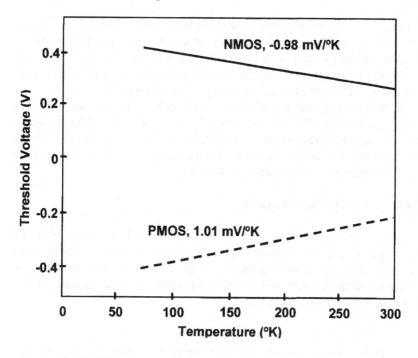

Figure 6-5. Variation of threshold voltage with temperature [8].

Where Q_{eff} is the effective oxide charge per unit area, ε_{Si} is the dielectric constant of silicon, V_{SB} is the source to body voltage, N is the channel surface concentration, and Φ_{ms} is the work function difference between gate and the channel. The Fermi potential of the bulk silicon is a temperature dependent parameter and as the temperature is reduced, Φ_B increases and therefore the threshold voltage V_{TH} is increased. Figure 6.5 shows the variation of the threshold voltage with temperature. In some applications the increase in threshold voltage is compensated with applying small forward body bias (200-400 mV) to source substrate junction [20]. The increased *pn*-junction current due to forward body bias is negligible in very low temperature because of the decrease in the intrinsic carrier concentration.

2.4   Leakage Currents and Short Channel Effects

An important improvement with the low temperature operation is the leakage current reduction. This can be illustrated using single transistor dynamic memory cell [21]. The length of the time for which the information can be stored in the capacitor is limited by the *pn* junction leakage, I_j, and the transistor off state leakage I_d. Figure 6.6 shows the behaviour of the drain current, I_d, as the temperature is reduced from 100°C to -50°C. The device has a leakage current of 10^{-8}A in 100°C at gate voltage of zero. The leakage of the device is reduced to 10^{-13}A for the gate voltage of zero at -50°C. This is a decrease in leakage current by five order of magnitude. The memory cell information retention time increases by the same factor.

In the following subsections the low temperature characteristic of leakage currents and short channel effects will be discussed. The leakage currents include the subthreshold current, drain induced leakage current (*GIDL*) and Impact ionization substrate current. The short channel effects include drain induced barrier lowering and punch through.

2.4.1   Subthreshold Leakage

When the gate voltage of the *MOS* transistor is reduced below the threshold voltage, the transistor enters into the off mode. For an ideal transistor, the drain current in the off mode is zero. However, in reality the drain current for the gate voltage less than the threshold voltage is non-zero and is called subthreshold current.

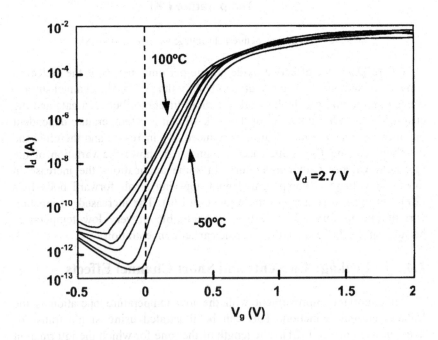

Figure 6-6. Drain current I_d, in a one device memory cell [21].

The *MOS* transistor drain current is given by Eq. (6.3), and under subthreshold regime $(V_G - V_{TH})$ becomes negative signifying the subthreshold region of its operation [8]:

$$I_D = I_o \exp\left(\frac{q(V_G - V_{TH})}{nkT}\right) \tag{6.3}$$

Where V_{TH} is the threshold voltage, k is the Boltzmann's constant, T is the absolute temperature, and n is a junction capacitance factor and its value does not change with temperature.

Figure 6.7 shows typical I_D–V_g characteristics for an n-channel *MOSFET* at different temperatures. As it is illustrated in the figure, the subthreshold slope (dashed lines) increases as temperature is decreased. These lines merge at a point which corresponds to threshold voltage near 0°K. The typical value of the subthreshold current in room temperature is 70-110 mV/decade. At 77°K the subthreshold slope increased by a factor of approximately 4.

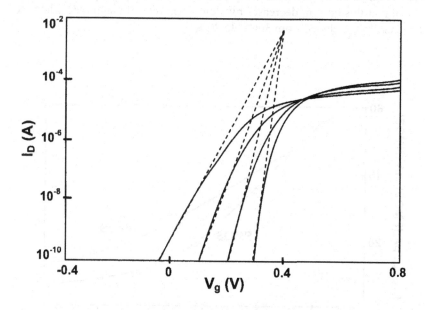

Figure 6-7. Steeping of subthreshold characteristics at low temperature [8].

2.4.2   Impact Ionization Substrate Current

Characterization of the substrate current due to the impact ionization is important to study the kink effect and hot carrier degradation processes. It has been found that the classical substrate current model is applicable down to near liquid helium temperature (20°K) [22]. Figure 6.8 shows the variation of the maximum substrate current with the temperature.

It can be seen in Figure 6.8 that the substrate current for both *NMOS* and *PMOS* transistors is increased when the temperature is decreased. However, the substrate current for *NMOS* device is higher than the substrate current for *PMOS* device by approximately 3 orders of magnitude. This is due to the higher mobility and higher rate of impact ionization of electrons.

2.4.3   Gate Induced Drain Leakage (*GIDL*)

Gate induced drain leakage current (*GIDL*) is one of the major concerns in scaled *MOS* technologies [23, 24]. The *GIDL* is mainly due to the band to band tunneling in the drain region at high drain voltage while biasing the device into strong accumulation. The *GIDL* strongly depends on the gate to drain overlap region and also on the drain doping level. The *GIDL* effect has been studied under different temperatures (20-300°K) [25] and has been shown that this current decreases by almost an order of magnitude when the temperature is reduced from 300°K to 20°K.

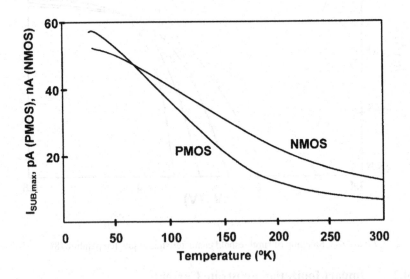

Figure 6-8. Variation of the maximum substrate current with temperature for *N* and *P* channel *MOS* devices (*W/L*=50/10μm, V_d=3 and -4.5 V) [22].

2.4.4   Drain Induced Barrier Lowering (*DIBL*)

Drain induced barrier lowering effect causes the reduction in threshold voltage when a voltage is applied to the drain of the *MOS* transistor such that [26, 27]:

$$V_{TH} = V_{TH0} - \lambda \cdot V_d \tag{6.4}$$

Where V_{TH} is the threshold voltage, V_{TH0} is the threshold voltage at zero drain bias, and λ is the *DIBL* coefficient. The λ is obtained by measuring the output current to gate voltage ratio. The variation of the λ is studied for different temperatures and it has been found to be temperature independent. This indicates that *DIBL* remains a serious constraint even under cryogenic condition since it can not be reduced by cooling down.

2.4.5   Punch Through

The punch through current represents the leakage current from the source to the drain of the *MOS* transistor via the silicon substrate and is given by:

$$I_d = I_0 \exp\left(-\frac{\Phi}{kT}\right) \tag{6.5}$$

2.5   Interconnect Resistance

I realistic VLSI circuits, interconnection between transistors play an important role in performance of the circuit. As the length of interconnects increases, their parasitic capacitances and resistances are also increased contributing to the propagation delay. In scaled geometries, due to increased transistor densities, the average interconnect length is increased. Moreover, its cross section area is reduced resulting in even higher delay in scaled technologies.

The low temperature operation of the circuits will offer tremendous advantage due to significant increase in the interconnect conductivity. For example the conductivity of aluminum is about on order of magnitude larger at 77°K than at 300°K [28, 29]. Modern *MOSFETs* are fabricated with junction and polysilicides to reduce resistance and to prevent metal penetration into shallow junctions. For degenerately polysilicon, the increase in conductivity is around 20% [28]. The conductivity of the *TiSi₂* is 3-4 times more at 77°K than at 300°K [30]. Table 6.1 summarizes variations in device and interconnect parameters [31.

Table 6-1. Normalized resistance variation with temperature in *CMOS* technologies [31].

Resistance	300°K	77°K	4.2°K
N⁺ Diffusion	1	0.76	0.72
Polysilicon	1	0.89	0.88
Aluminum	1	0.14	0.05
P-well	1	0.30	$< 10^{-5}$

3.   RELIABILITY AT LOW TEMPERATURE

Lowering the chip temperature is expected to improve the reliability of the overall system. Nearly all degradation mechanisms in electronic devices, such as electromigration, inter-diffusion, and corrosion have a thermal activation component that follows the Arrhenius relationship. Since the rate of degradation decreases exponentially with decreasing the temperature, orders of magnitude improvement in reliability could be expected with cooling.

The mean time to failure (*MTTF*) for thermally activated mechanisms is proportional to a temperature dependent term expressed by the Arrhenius relation as follow [8]:

$$MTTF \propto \exp\left[\left(\frac{E_a}{k}\right)\left(\frac{1}{T_0} - \frac{1}{T_R}\right)\right] \qquad ($$

$$MTTF \propto \exp\left[\left(\frac{E_a}{k}\right)\left(\frac{1}{T_0} - \frac{1}{T_R}\right)\right] \qquad (6.6)$$

Where T_0 is the operating temperature and T_R is the reference temperatures. E_a is a fitted parameter related to the activation energy of a given thermally activated process and is in the range of 0.3eV to 1.2eV. Based on Eq. (6.5) and assuming $E_a = 0.3eV$, operating at 77°K will improve the *MTTF* by approximately 5 orders of magnitude compared to operating at 300°K [8].

3.1   Electromigration

Electromigration is the creation of metal voids and shorts in interconnects due to movement of metal atoms at high current densities. The commonly used model to predict the *MTTF* caused by electromigration assumes a thermal process with activation energy of 0.7eV. Electromigration decreases with temperature and at 77°K the *MTTF* due to electromigration improves significantly [8].

3.2   Hot Carriers

Electrons and holes in the inversion layer gain kinetic energy from the electric field and loose this energy to lattice collision. If the electric field is high enough, the carrier energy exceeds the lattice thermal energy and therefore carrier is called hot carrier. A fraction of these hot carriers are injected into the gate oxide and contribute to the gate leakage. However a small fraction of the injected carriers becomes trapped in the gate oxide and alters the *MOSFET* threshold voltage. As the temperature is reduced, the carriers mean free path increases due to the reduction in thermally generated lattice vibrations. This results in larger fraction of carriers reaching the gate oxide therefore higher susceptibility to hot carrier degradation at low temperature. Some studies have suggested that the hot carrier degradation in low temperature (77°K) is due to increased influence of the trapped charge on the device operation [32].

More recent studies have shown that for 100nm technology and beyond (lower V_{DD} and thinner gate oxide) the temperature dependence of hot carriers is reversed because carriers with sufficient energy lie only in the tail of the energy distribution [33]. When the temperature is reduced, the number of carriers with enough energy for impact ionization decreases. This reduces I_{SUB} and hot carrier degradation. Figure 6.9 shows the percentage of the I_{dsat} degradation of a 50nm *NMOS* transistor after hot carrier stressing. It can be seen from this figure that I_{dsat} degradation is improved at -50°C.

4.   MICROPROCESSOR LOW TEMPERATURE OPERATION: A CASE STUDY

Potential advantages of using refrigeration for cooling processors have been reported in the past [34]. Low temperature operation can reduce important device scaling and circuit performance barriers in sub-130nm *CMOS* technologies. It permits the scaling of supply voltages of high speed circuits to sub-1V by reducing the sub-threshold currents and increasing the carrier mobility in the channels, lowering interconnection resistances significantly, and reducing interconnection related failure mechanisms. In this section, tradeoffs in microprocessor clock frequency, energy efficiency (*MIPS/Watt*), die area and system power are described when active cooling is used to reduce the operating junction temperature of the microprocessors below a typical hot temperature of 90°C.

The purpose of this work was to find out if low temperature *CMOS* operation has any merit for scaled technologies where transistor subthreshold leakage is relatively high. And if yes, what kind of device, circuit, and design choices are applicable for high performance microprocessors [35]. Consequently, the above mentioned tradeoffs were studied by combining active cooling with:

- Supply voltage (V_{DD}) selection.
- Applying body bias.
- Sizing of transistors in critical and non-critical paths on chip.
- Reduction of channel length (*L*) as a function of different process technology worst case leakage limits.

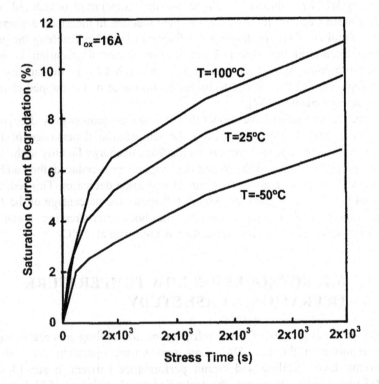

Figure 6-9. NMOSFET I_{dsat} degradation after hot carrier stressing at $V_d=V_g=2V$ [33].

Active cooling with refrigeration and without refrigeration was considered. Several active cooling techniques including air cooling, liquid cooling and refrigeration were investigated. Refrigeration is the most effective cooling solution and is considered for junction temperatures not much below the ambient temperature. Cooling power was considered as part of total system power tradeoffs. System power is the total of chip power (switching and leakage) and power consumed by the cooling system. Analytical models are used for frequency, power, die area, etc. in an electro-thermal analysis tool. The tool analyzes the followings:

- Frequency, limited by logic and interconnect *RC* paths.
- System energy efficiency.
- Chip switching and leakage powers, including subthreshold and gate oxide leakage.
- Package and cooling system characteristics.
- Die area.
- Gate oxide reliability-limited maximum V_{DD} constraints.
- Maximum temperature in a self-consistent manner.

The model parameters and input parameters to the tool are typical values. The parameters are extracted from device measurements, process files, and chip measurements.

4.1 High Leakage vs. Low Leakage

Figure 6.10 shows the frequency and power tradeoffs for iso-reliability high performance operation and iso-power operation conditions when refrigeration active cooling is incorporated. The relative contributions of cooling power, dynamic power, and leakage power demonstrate how leakage power and cooling power can be traded off. This is best shown in the iso-power case. For a constant power limit of 80W, the frequency increases by 4.5% going from air cooling to refrigeration in a low-leakage technology, and by 7.5% for high- leakage technology (Figure 6.10). This happens because when leakage is a large percentage of the total power (31% in this case), the leakage power reduction due to lower T_j translates to higher savings in total chip power. Then, power overhead of the cooling system will have less impact on total system power. To achieve the highest operating frequency in line with microprocessor applications, the iso-reliability case must be studied as shown in Figure 6.10.

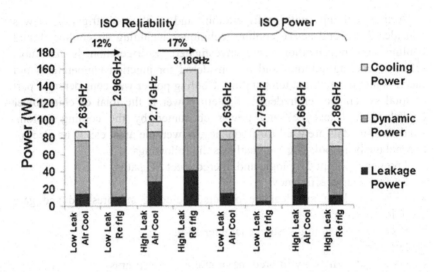

Figure 6-10. Reliability and power limited maximum frequency achievable for low and high leakage technologies with refrigeration.

On the other hand, for a low-leakage technology, the reliability-limited frequency improves by 12% and the system power increases by 35% going from air cooling to refrigeration. When leakage is higher, the frequency increases by 17% for a 62% increase in system power. Therefore, the frequency vs. power tradeoff is worse when the leakage is higher. Frequency improvements in both cases come from operation at reduced temperature and the higher V_{DD} allowed at lower T_j due to utilizing of refrigeration.

4.2   Optimal Design at Low Temperature: Power, Frequency and Energy Comparison

Now that the active cooling for different amounts of worst-case process technology leakage constraints have been studied, the optimum design for low temperature *CMOS* operation can be investigated. The following design techniques for optimal low temperature operation are considered: optimizing V_{DD}, transistor channel length, transistor sizing, enhancing the process technology, and applying body bias. Figure 6.11 shows tradeoffs in system power, energy efficiency and die area vs. frequency offered by forward body bias (*FBB*), shortening L, changing V_{DD} and transistor sizing, with and without refrigeration.

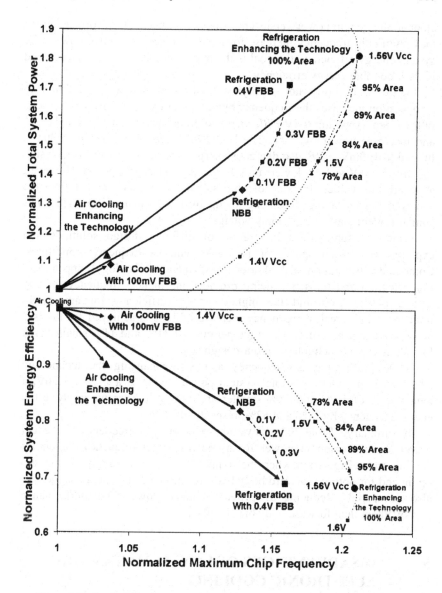

Figure 6-11. Tradeoffs in system power, energy efficiency, die area, and frequency by different circuit and design techniques.

System power and system energy efficiency as a function of chip frequency is plotted in Figure 6.11. These graphs are normalized to air cooling power, energy efficiency and frequency. When refrigeration combined with a design technique is utilized, the goal is to minimize the slope in the system power versus chip frequency curves. This corresponds to maximum chip frequency increase for lowest increase in system power. For system energy efficiency, the goal is the maximum change in frequency and highest possible energy efficiency.

Figure 6.11 shows how applying forward body bias in addition to refrigeration increases the frequency but the rate of system power increase is rather steep. Applying 0.4V *FBB* increases frequency by an additional 2.7% and increases power by 27%. The best *FBB* tradeoff is when its value is limited to 100mV. *FBB* also degrades energy efficiency by 16%. Decreasing V_{DD} from 1.56V to 1.4V lowers both frequency and system power. However, at lower V_{DD} values, the rate of chip slowdown is much higher than the achieved power saving. Reducing sizing by lowering transistor width has similar tradeoffs as for the supply voltage.

Table 6.2 summarizes integration of different design solutions and explores the design space for iso-power and iso-frequency conditions. Combined refrigeration with shorter L (enhancing technology), appropriate V_{DD} selection and transistor sizing provides the highest frequency for any system power limit and the highest energy efficiency for any target frequency. The greatest frequency increase of 11% is achieved for the iso-power case at a V_{DD} of 1.41V, a temperature of 31°C and 11% smaller area for enhancing the technology in our design space.

While performing iso-frequency analysis, enhancing the technology (shorter L), provides 38% total system power saving at a V_{DD} of 1.36V, a temperature of 15°C and 33% smaller area. In both cases we improve the energy efficiency by 11% and 62%, respectively.

In summary, Table 6.2 shows improvements in frequency for equal power and reduction in power for a specific target frequency for air air cooling and refrigeration when the transistor sizing and supply voltage are optimized for optimal forward body bias and shorter L. Die area changes are also compared. Reducing L provides better power and performance improvement than forward body bias in all cases.

5.   DISADVANTAGES OF LOW TEMPERATURE ELECTRONIC COOLING

There are two major disadvantages in utilization of cooling hardware in low temperature operation of electronic devices. One of these disadvantages is the cost of cooling hardware and the other disadvantage is the power consumption of cooling hardware. However, these disadvantages must be weighed against the advantages such as performance enhancement, leakage reduction, and reliability improvement of the electronic devices.

Table 6-2. Optimum design space for active cooling at iso-power and iso-frequency conditions in an ambient temperature of 35°C.

Iso Power	V_{DD} (V)	T_i (°C)	Area	Leakage Power	Cooling Power	Energy Efficiency	Frequency
Air Cooling Reference	1.49	80	100%	19%	14%	100%	100%
Air Cooling (100mvFBB)	1.47	81	95%	21%	14%	102%	102%
Air Cooling Enhanced Technology	1.40	82	89%	31%	14%	107%	107%
Refrigeration (200mV FBB)	1.60	30	73%	13%	29%	108%	108%
Refrigeration Enhanced Technology	1.41	31	89%	18%	28%	111%	111%
Iso Frequency	V_{DD} (V)	Tj (°C)	Area	Leakage Power	Cooling Power	Energy Efficiency	Frequency
Air Cooling Reference	1.49	80	100%	19%	14%	100%	100%
Air Cooling (100mvFBB)	1.45	76	89%	19%	14%	115%	87%
Air Cooling Enhanced Technology	1.46	20	73%	8%	29%	134%	75%
Refrigeration (200mV FBB)	1.34	72	78%	27%	14%	137%	73%
Refrigeration Enhanced Technology	1.36	15	67%	12%	30%	162%	62%

Adding cooling hardware to the electronics adds cost to the system. The cooling cost of a workstation microprocessor with non-redundant cooling hardware is 10% to 20% of the total cost of the system [1]. For a large server with redundant cooling hardware the cooling cost is higher in dollars, but it is smaller percentage of the total cost due to high overall cost of the system.

The cooling hardware consumes extra power during the startup and during the normal operation. The highest power is drawn for refrigeration during the compressor startup. The startup current draw of an *AC* single phase compressor motor can be five to seven times higher than the operating current, causing strain on the power system. Even for the normal operating condition power consumption of the cooling hardware can be a considerable percentage of the total system power.

Considering disadvantages of low temperature operation, there is no advantage to adding low temperature cooling hardware unless the chip module can not be maintained at acceptable temperatures with another cooling technique such as air cooling.

6. COOLING TECHNOLOGIES

For many years cooling technologies have played a key role in enabling and facilitating the packaging and performance improvements while satisfying reliability objectives in each new generation of computers. In this section we briefly present the work that Chu et. al. have conducted as a review of cooling technologies for computer systems [36].

6.1 Internal Module Cooling

The heat primarily is transferred through conduction and is internal to the chip module. The conduction rate depends on the thermal resistance of the module. The thermal resistance in turn depends on the physical structure and material properties of the module. The objective is to transfer the heat from electronic circuits to outer surface of the module where the heat will be removed by external means. As chip power levels continue to increase, different companies employ more conductive thermal interface materials (*TIM*) to meat the thermal resistance requirements. They also utilize more effective heat spreaders to move the heat from hot spots where the heat flux is 2 to3 times higher than average chip heat flux [37].

6.2 External Module Cooling

Cooling external to the module serves as the primary means to effectively transfer the heat generated within the module to the system environment by attaching a heat sink to the module. Traditionally the system environment has been air because of its simplicity, lower cost, and transparency to the customer. Beside the air cooled heat sinks, there are also liquid cooled heat sinks typically referred to as cold plates.

Fins

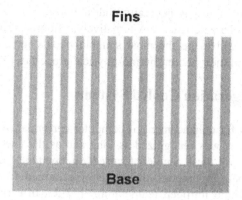

Base

Figure 6-12. Typical air cooled heat sink [36].

A typical air cooled heat sink is shown in Figure 6.12. The heat sink is constructed of a base which is in contact with the module and fins extended from the base to provide the heat transfer to the air. The heat is conducted through the base and into the fins and then to the air flowing in the spaces between fins by convection. The parameters that influence the thermal performance of the heat sink include the thickness and the plan area of the base; fin thickness, height, spacing, and surface area; and material thermal conductivity. Many studies have been conducted to optimize the external thermal resistance for particular application conditions [38-40].

For modules with higher power consumption where air cooling can not meet the thermal requirements, heat is removed from the modules using water cooled cold plates. In this technique the plates in contact with the module are cooled using the water. Compared to air cooling, water cooling can provide almost an order of magnitude reduction in thermal resistance due to the higher thermal conductivity of the water. In order to extend the efficiency of the water cooled cold plates, researchers have developed microchannel cooling structure where water is conducted into the chip package through chemically etched deep microchannels [41].

Immersion cooling is another method to cool down the high heat flux components. Unlike the water cooled cold plate method which physically separates the coolant from the chip, immersion cooling brings the coolant in direct contact with the chip. Direct liquid immersion cooling offers a high heat transfer coefficient which reduces the temperature rise of the heated chip surface above the liquid coolant temperature. There are several coolants that can provide adequate cooling but only few of them will be chemically compatible. For example, water is a liquid which has very desirable heat transfer properties, but it is generally undesirable due to its poor chemical and electrical characteristics. Alternatively, fluorocarbon liquids are generally considered to be the most suitable liquids for direct immersion cooling, in spite of their poorer thermo-physical properties [42, 43].

6.3   Refrigeration Cooled Systems

The potential for enhancement of computer performance by operating at lower temperatures was recognized as long ago as the late sixties and mid seventies. Some of these studies focused on operating at liquid helium temperature (4°K). In nineties the focus was shifted to *CMOS* devices operating near liquid nitrogen temperatures (77°K). In early nineties, IBM initiated an effort to demonstrate the feasibility of the packaging and cooling of a *CMOS* processor in a form suitable for product use [44]. A major part of the effort was devoted to the development of a refrigeration system that would meet IBM's reliability and life expectancy specifications and handle a cooling load of 250W at 77°K. As a result of this effort, prototype *Stirling Cycle* Cryo-coolers compatible with overall system packaging constraints were built and successfully tested. IBM's most recent interest in refrigeration cooling is focused on the application of *Conventional Vapour Compression* refrigeration technology to operate below room temperature conditions, but well above cryogenic temperatures.

In 1997, IBM developed and shipped its first refrigeration cooled server (S/390 G4 system) [1, 45]. This cooling technique provided an average junction temperature of 40°C which amounted to temperature decease of 35°C below that of an air cooled system. The system packaging layout is shown in Figure 6.13. In this figure, below the bulk power compartment is the central electronic complex (*CEC*) where the multi chip module (*MCM*), housing twelve processors, is located. Two modular refrigeration units (*MRU*) located near the middle of the frame provide cooling via the evaporator attached to the back of the processor module. Only one *MRU* is operated at a time during normal operation. The evaporator mounted on the processor module is fully redundant with two independent refrigerated passages. Refrigerant passing through one passage is adequate to cool the *MCM* which dissipates a maximum power of 1050W.

Following the success of this machine, IBM and other companies such as Fujitsu [46] has continued to exploit the advantage of sub-ambient cooling at high-end product line.

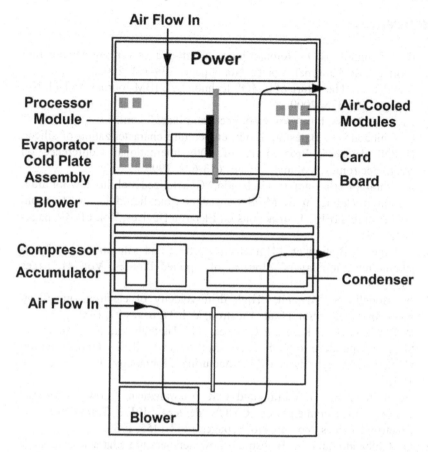

Figure 6-13. IBM S390 G4 server with refrigeration-cooled processor module and redundant modular refrigeration units [1, 45].

7.   SUMMARY

The advantages of low temperature operation of electronics and CMOS devices have been studied by many researchers. In this chapter the low temperature operation of the CMOS devices was reviewed and advantages and disadvantages of utilization of the refrigeration cooling was presented. Finally the cooling technologies were briefly described.

References

1. R.R. Schmidt, B.D. Notohardjono, "High-end server low-temperature cooling", IBM Journal, Vol 46, No. 6, pages 739-751, 2002.
2. R.D. Isaac, "The future of CMOS technology", IBM Journal, Vol 44, No. 3, pages 369-378, 2000.
3. Kryotech News; http://www.kryotech.com/index2.html.
4. G. Ghibaudo, F. Balestra, "Low temperature characterization of silicon CMOS devices", Proceedings of 20th International Conference on Microelectronics, Volume 2, Pages 613-622, 1995.
5. A. Emrani, F. Balestra, G. Ghibaudo, "Generalized mobility law for drain current modeling in Si MOS transistors from liquid helium to room temperature", IEEE Transactions on Electron Devices, Vol. ED-40, pages 564, 1993.
6. F. Fang, A.B. Fowler, "Hot electron effect and saturation velocity in silicon inversion layers", Journal of Applied Physics, Vol. 41, pages 1825, 1970.
7. A. Modelli, S. Manzini, "High drift velocity of electrons in silicon inversion layers", Solid State Electron, Vol. 31, pages 99, 1988.
8. W.F. Clark, B. El-Kareh, R.G. Pires, S.L. Titcomb, and R.L. Anderson, "Low temperature CMOS - A brief review", IEEE Transaction on Components, Hybrids, and Manufacturing Technology, Vol. 15, No. 3, 1992.
9. A. Hairapetian, D. Gitlin, and C.R. Viswanathan, "Low-temperature mobility measurements on CMOS devices" IEEE Transactions on Electron Devices, Vol. 36, No. 8, pages 1448-1455, 1989.
10. G. Ghibaudo and F. Balestra, "Low temperature characterization of silicon CMOS devices", in Proceedings of the 20th International Conference on Microelectronics, 1995.
11. S.B. Broadbent, "CMOS operation below freezeout", in Proceedings of the Workshop on Low Temperature Semiconductor Electronics, 1986.
12. G. Gildenblat, L. Colonna-Romano, D. Lau, and D. E. Nelsen, "Investigation of cryogenic CMOS performance", in Technical Digest, International Electron Devices Meeting, 1985.
13. K.J. Hass, H.C. Shaw, "Cryogenic operation of ultra low power CMOS", 9th NASA Symposium on VLSI Design, pages 3.4.1-3.4.7, 2000.
14. I.M. Hafez, G. Ghibaudo, F. Balestra, M. Haond, "Impact of LDD structures on the operation of silicon MOSFETs at low temperature", Solid State Electron, Vol. 38, pages 419, 1995.
15. F. Balestra, L. Audaire, C. Lucas, "Iilfluence of substrate freeze-out on the characteristics of MOS transistors at very low temperature", Solid State Electron, Vol. 30, pages 321, 1987.

16. E. Simoen, B. Dierickx, L. Warmerdam, J. Vermeiren, C. Claeys, "Freeze-out effects on NMOS transistor characteristics at 4.2K", IEEE Transaction on Electron Devices, Vol. ED-356 pages 1155, 1989.

17. I.M. Hafez, G. Ghibaudo, F. Balestra, "Reduction of kink effect in short channel MOS transistors", IEEE Electron Device Letter, VGI. EDL-11, pages 120, 1990.

18. LM. Hafez, G. Ghibaudo, F. Balestra, "Analysis of the kink effect in MOS transistors", IEEE Transaction on Electron Devices, Vol. ED-37, pages 818, 1990.

19. D. Foty, "Impurity ionization in MOSFETs at very low temperature", Cryogenics, Vol. 30, pages 1056, 1990.

20. R.L. Anderson, J.L. Hill, "Low temperature electronics", Micro-electron Journal, Vol. 19, pages 7-12, 1988.

21. A. Keshavarzi and C.F. Hawkins, "Intrinsic Leakage in Low Power Deep Submicron CMOS ICs", Proceedings of the IEEE International Test Conference, pages 146-155, 1997.

22. K. Rais, G. Ghibaudo, F. Balestra, "Temperature dependence of substrate current in silicon CMOS devices", Electron Letters, Vol. 29, pages 778, 1993.

23. J. Chen, T.Y. Chan, I.C. Chen, P.K. KO, C. Hu, "Sub breakdown drain leakage current in MOSFET", IEEE Electron Device Letters, Vol. EDL-8, pages 515, 1987.

24. K. Kurimoto, Y. Odake, S. Odanak, "Drain leakage current characteristics due to band to band tunneling in LDD MOS devices", IEDM Technical Digest, pages 621, 1989.

25. K. Rais, F. Balestra. G. Ghibaudo, "Temperature dependence of gate induced drain leakage in silicon MOS devices", Electron Letters, pages 32, 1994.

26. T. Grotjohn, B. Hoefflinger, "A parametric short channel MOS transistor model for subthreshold and strong inversion current" IEEE Transaction on Electron Devices, Vol. ED-31, pages 234, 1984.

27. S. Chamberlain, S. Ramanan, "Drain induced barrier lowering analysis in VLSI MOSFET devices using two dimensional numerical simulations", IEEE Transaction on Electron Devices, Vol. ED-33, pages 1745, 1986.

28. F.H. Gaensslen, L. Rideout, E.J. Walker, J.J. Walker, "Very small MOSFETs for low temperature operation", IEEE Transactions on Electron Devices, Vol.ED-24, pages 218-229, 1977.

29. R.K. Kirschman, "Cold electronics: An overview", Cryogenics, Vol. 25, pages 115-122, 1985.

30. L. Krusin-Elbaum, J.Y.C. Sun, C.Y. Ting, "On the resistivity of TiSi2: The implication for low temperature applications", IEEE Transaction on Electron Devices, Vol. ED-34 Pages 58-62, 1987.

31. H. Hanamura, M. Aoki, T. Masuhara, O. Minato, Y. Sakai, T. Hayashida, "Operation of bulk CMOS devices at very low temperature", IEEE Journal of Solid State Circuits, Vol. SC-21, No. 3, pages 484-490, 1986.

32. P. Heremans, G. Van den Bosch, R. Bellens, G. Groesencken, H. Maes, "The dependence of channel hot carrier degradation on temperature in the range 77K to 300K", Proceedings of 19th ESSDERC, pages 727-731, 1989.

33. Y. Bin, W. Haihong, C. Riccobene, K. Hyeon-Seag, X. Qi, L. Ming-Ren, C. Leland, H. Chenming, "Nanoscale CMOS at low temperature: design, reliability, and scaling trend", Proceedings of Technical Papers, International Symposium on VLSI Technology, Systems, and Applications, pages 23-25, 2001.

34. I. Aller, K. Ghoshal, H. Schettler, S. Schuster, Y. Taur, and D. Torreiter, "CMOS Circuit Technology for Sub-Ambient Temperature Operation", IEEE International Solid-State Circuits Conference, pages 214-215, 2000.

35. A. Vassighi, A. Keshavarzi, S. Narendra, G. Schrom, Y. Ye, S. Lee, G. Chrysler, M. Sachdev, V. De, "Design optimizations for microprocessors at low temperature", Proceedings of Design Automation Conference, pages 2-5, 2004.

36. R.C. Chu, R.E. Simons, M.J. Ellsworth, R.R. Schmidt, and V. Cozzolino, "Review of Cooling Technologies for Computer Products", IEEE Transaction on Device and Materials Reliability, Vol. 4, No. 4, pages 568-585, 2004.

37. J.U. Knickerbocker, "An advanced multichip module (MCM) for highperformance unix servers", IBM Journal, vol. 46, No. 6, pages 779–804, 2002.

38. D.J. De Kock and J.A. Visser, "Optimal heat sink design using mathematical optimization", Advance Electronic Packaging, Vol. 1, pages 337–347, 2001.

39. J.R. Culham and Y.S. Muzychka, "Optimization of plate fin heat sinks using entropy generation minimization", IEEE Transaction on Component Packaging Technology, Vol. 24, No. 2, pages 159-165, 2001.

40. M.F. Holahan, "Fins, fans, and form: Volumetric limits to air-side heat sink performance", Proceedings of 9th Intersociety Conference Thermal and Thermomechanical Phenomena in Electronic Systems, pages 564-570, 2004.

41. D.B. Tuckerman and R. F. Pease, "High performance heat sinking for VLSI", IEEE Electron Device Letters, Vol. EDL-2, No. 5, pages 126–129, 1981.

42. A.E. Bergles and A. Bar-Cohen, "Direct liquid cooling of microelectronic components", Advances in Thermal Modeling of

Electronic Components and Systems, A. Bar-Cohen and A. D. Kraus, Eds. New York: ASME Press, Vol. 2, pages 233-342, 1990.

43. R.E. Simons, "Direct liquid immersion cooling for high power density microelectronics", Electronic Cooling, Vol. 2, No. 2, 1996.

44. R.E. Schwall and W.S. Harris, "Packaging and cooling of low temperature electronics", Advances in Cryogenic Engineering, New York: Plenum Press, pages 587-596, 1991.

45. R.R. Schmidt, "Low temperature electronics cooling", Electronics Cooling, Vol. 6, No. 3, 2000.

46. A. Fujisaki, M. Suzuki, and H. Yamamoto, "Packaging technology for high performance CMOS server fujitsu GS8900", IEEE Transaction on Advanced Packaging Vol. 24, pages 464-469, 2001.

Index